Copper Overview - From Historical Aspects to Applications

Edited by Daniel Fernández González

Published in London, United Kingdom

Copper Overview - From Historical Aspects to Applications
http://dx.doi.org/10.5772/intechopen.1001615
Edited by Daniel Fernández González

Contributors
Ahlam I. Al-Sulami, Ahmad Poursattar Marjani, Ahmed Elkhatat, Bawindsom Marcel Kébré, Berit V. Eriksen, Betty Revon Liu, Daouda Oubda, Fatemeh Sadegh, François Zougmoré, Hadi Sharifidarabad, Iman Mohammadi Dehcheshmeh, Jonathan Wijaya, Khurram Saleem, Lorenz Kienle, Mei Wang, Meng-Jiun Lai, Mohammad Ebrahim Soltani, Moiken Hinrichs, Shengjia Li, Soumaïla Ouédraogo, Tesfay G. Ashebr, Ulchitsky Oleg Alexandrovich, Xinguo Ma, Yue-Wern Huang, Zacharie Koalga, Zhengwang Cheng

First published in London, United Kingdom, 2024 by IntechOpen
IntechOpen is the global imprint of INTECHOPEN LIMITED, registered in England and Wales, registration number: 11086078, 167-169 Great Portland Street, London, W1W 5PF, United Kingdom

British Library Cataloguing-in-Publication Data
A catalogue record for this book is available from the British Library

Additional hard and PDF copies can be obtained from orders@intechopen.com

Copper Overview - From Historical Aspects to Applications
Edited by Daniel Fernández González
p. cm.
Print ISBN 978-0-85466-158-9
Online ISBN 978-0-85466-157-2
eBook (PDF) ISBN 978-0-85466-159-6

For EU product safety concerns:
IN TECH d.o.o., Prolaz Marije Krucifikse Kozulić 3, 51000 Rijeka, Croatia,
info@intechopen.com or visit our website at intechopen.com.

Meet the editor

Daniel Fernández-González graduated with a bachelor's and master's degree in mining engineering in 2013, and he earned a Ph.D. in materials from the Oviedo School of Mines, Energy, and Materials at the University of Oviedo, Spain, in 2019. He has worked as a researcher at the Nanomaterials and Nanotechnology Research Center (CINN-CSIC) since 2021. His research deals with the synthesis of metals, ceramics, and composite materials and with sustainable metallurgical processes based on solar energy. He has partcipated in diferent research projects funded by public agencies and private companies. He has published more than 40 artcles and six books. He has been a visitng researcherat the Universidad Panamericana (Mexico), the AGH University (Poland), and the PROMES-CNRS laboratory (France).

Contents

Preface

Copper is a metal that is inextricably associated with the history of humanity, as witnessed by the historical periods of the Copper Age and Bronze Age, whose names are derived from the importance that the metal and the alloy have had in those periods of human history. In this way, since its discovery as a native metal, copper has found application in numerous fields, either in an ultrapure state or alloyed with other metals. The importance of copper, and its utilization in different applications, derives from its remarkable properties: high ductility, malleability, thermal conductivity, electrical conductivity, and corrosion resistance, among others that are still being discovered in advanced fields. Nowadays, copper is used for electrical purposes (power transmission and generation, building wiring, electrical and electronic products, etc.) with approximately 75% of the metal produced in the world used in this field. At this point, it is important to highlight the significance of copper in the energy transition. For example, 2.5–6.4 tons per MW are employed in a wind turbine. This field, even when it is related to the electrical purposes previously reported, has represented an important increase in the demand for copper in the last decades. The electric vehicle has also had a key impact on the copper demand. Therefore, copper is the third most-produced metal in the world after iron and aluminum, and one of the most-produced materials in general. The refinery production is estimated at 27 Mt in 2023 (22 Mt mine production), while the production in 1994 had reached 9.5 Mt (mine production). This data displays the still growing importance of the metal for humans and also the greater and greater importance of recycling. The sustainability and recycling topics, although they are not deeply treated in this book, might be a subject of interest for newly edited books from IntechOpen. Anyway, volumes of production are important from the traditional point of view. Nevertheless, in other fields still being researched, copper (and copper compound) applications will have more of an economic impact than simply importance in terms of quantity. This is, for instance, the case of the potential application of copper and copper-based nanoparticles in medicine. Thus, it is possible to see that copper still has importance as a subject of research and in economic terms. The number of copper applications is expected to grow and expand to other new uses, probably many of them still to be identified. Therefore, investigation into copper and copper materials will be an important area in the next years.

This book has been written by expert researchers from different countries, which gives this book an international vision. The book is organized into several sections dedicated to various aspects of copper metallurgy, from the historical aspects to the new applications of the metal in advanced fields and recycling possibilities. The publication of this book has involved several people, from the authors to the publishing company, and for that reason, I want to thank everyone for their support in the successful conclusion of this project. Besides, I want to thank the authors for their outstanding contributions, willingness to share their research and experience with the community, and intense work in making this book a reality. I also want to express my gratitude to my colleagues in the Nanomaterials and Nanotechnology Research Center (CINN-CSIC) and the University of Oviedo (particularly Professor Luis Felipe Verdeja) for their support during the elaboration of this book. Finally, I want to thank IntechOpen for the initiative to publish edited books, in general, and this book about copper, in particular, because edited books are always a good manner to share knowledge from a multidisciplinary point of view.

Daniel Fernández-González
Center for Research in Nanomaterials and Nanotechnology-Spanish, National Research Council (CINN-CSIC), El Entrego, Asturias, Spain

Section 1

Historical Aspects

Chapter 1

A New Typology of the Southern Urals Fortified Settlements of the Bronze Age Development

Ulchitsky Oleg Alexandrovich

Abstract

This chapter describes the results of a scientific study on the typology of the Southern Urals fortified settlements of the Bronze Age development. Archeological data comprehensive analysis are related to the architectural and urban planning interpretation of previously studied materials, using graphic restoration of architecture methods, which allowed us to clarify the definition of the settlements typology belonging to the ancient cultures of the Southern Urals. The dynamics of the basic typological criteria development, whereby it is proposed to determine a particular type of settlement, is taken into account. Four basic criteria introduce into a new typology of fortified settlements: space-planning structure of the layout nature, the size of settlements with account of surroundings, the morphology in the form of the plan, and fortification power level. In the train of the study, a table was compiled that shows the genesis of fortified settlements of the Bronze Age on the Southern Urals territory. Recommendations for the use of the typology in further research are given, in respect to the visualization of system connections between settlement objects. The method of graphic coding in the typological model development is used as the most effective for displaying informative data on diagrams and graphs.

Keywords: new typology, fortified settlements, the southern Urals, the bronze age, planning structures

1. Introduction

The Southern Urals of the Bronze Age fortified settlement typology is under development on the previous results basis but has a number of significant differences that allow identifying and classifying the objects according to the identified typological criteria more precise. Previously, three main types were distinguished: the *early Sintashta*, oval in plan shape, was a complex fortification system that had an internal and external citadel surrounded by an external fortification system—embankment and moats—and had a centrally planned structure; the "*classic*" *Sintashta*, round in plan shape, had a similar complex fortification; a third type rectangular *Petrovsky or Alakul* structure has been noted, closely resembling a square [1] or diamond shape in plan, which had a simplistic fortification, preserving the Sintashta planning pattern traditions [2], subsequently, this type was transformed into a linearly regular plan

IntechOpen

shape of the Sargarin-Alekseevskoye settlements [3]. The first thing that distinguishes this development from the previous ones is a universal approach to the shape of the primeval layer settlement plan analysis [4–7]. Only the first layer of the fortified settlement is taken as the basis for determining the plan shape, due to the fact that subsequent layers significantly deform the plan structure and thus complicate the original planning structure determination. The second significant difference of the proposed typology is the assessment of a fortified settlement fortification power level by three degrees.

Archeological data are updating and interpreting the process of local exploration using virtual reconstructions and modern methods of geo-survey and new data analysis. According to the recent study results, the archeological data interpretation has a vast variety and arouses research interest in terms of generating new scholarly knowledge on early forms of architecture, archaic building technologies, and settlement systems [8–10].

The purpose of the study is to develop an architectural planning typology of the Southern Urals fortified settlements of the Bronze Age.

This research has the following objectives:

- To analyze developments on the research topic and consider a number of examples on deciphering the fortified settlements' planning structures;

- To determine the main typological assessment criteria for the Southern Urals fortified settlements of the Bronze Age;

- To characterize and update a number of criteria for typological assessment using archeological data graphical interpretation;

- To develop and prove a criterion for the fortification strength assessment;

- To develop a diagram of the fortified settlements typology;

- To provide recommendations on the developed typology implementation and represent prospects related to the development results approbation.

As a result of the study, the new typology has been clearly investigated, enabling to analyze the fortified settlements' development dynamics on the territory of the Southern Urals in ancient times and build territorial system connections between different types of settlements. A diagram has been developed, clearly demonstrating the genesis of the studied objects according to the main criteria; in particular, criteria for the fortification strength assessment have been developed in detail.

The first criterion in determining the typology of the objects under study is *the space-planning pattern by the planning nature* (by reconstruction materials) [11].

Conclusions about the nature of the oldest Southern Urals fortified settlements' space-planning pattern were made at the early stages of their reconstruction. For more than two decades, the conclusions have been updated and described in detail in the works of historians and archeologists, art historians, and architects [12]. As a result of a planning structure and virtual modeling detailed architectural and urban planning analysis, two specific varieties of fortified settlements that had existed on the territory of the Ural-Tobolsk plateau were distinguished. The first type has conditionally been designated as fortified settlements "with an inhabitable walls pattern;"

the second type has been designated "with a continuous building pattern," by analogy with the definition introduced in the evaluation of the ancient Chorasmian architectural culture by S.P. Tolstov ([13], p. 10).

The second criterion for the development of a typological line is *the fortified settlement morphology according to the plan shape and the fortification strength*.

Earlier on, it was concluded that, mainly, fortified settlements with inhabitable walls, the settlements in which dwellings were closely adjacent to defensive walls and made up a whole with fortification, belong to the Sintashta period. The architecture of such structures may be divided into subtypes: the 1st subtype is small, and the 2nd subtype is large. Conditionally, fortified settlements with a total area of up to 6 hectares may be called "small" and settlements with an area of 6 or more hectares "large," taking into account the adjacent territory.

The classification of ancient Mesopotamian settlements of the Bronze Age according to Adams-Nissen is used in determining the type and sizes of Sintashta settlements. A three-level hierarchy of settlements in the Uruk region (Mesopotamia), formed during the Jemdet Nasr period (XXXI–XXIX centuries BC) and the Early Dynastic I (XXVIII–XXVII centuries BC), chronologically, corresponding to the nascent stage of the Sintashta fortified settlements in the Southern Urals, is given as an example. According to the settlement type classification, by the III millennium BC, 3–4 hierarchical types were formed in the Uruk region: Type 1 was considered small villages ("vilages") with an area of 0.1–6 hectares (about 124 settlements), "towns" ("towns") with an area of 6.1–25 hectares (21 settlements), and "central settlements" ("urban centers" with an area of more than 50 hectares (there were only two settlements), and the fourth level was a "city" formation with an area of more than 400 hectares ([14], p. 18). Another important factor in determining the settlement type classification, according to Adams-Nissen, is the distance between individual settlements and the formed separate groups.

Generally, one of the shortcomings of this methodology is a common approach to the analysis of build-up area indicators and transit distances between settlements. Although for many historical cities, the build-up area size was not a decisive indicator for determining whether a settlement belonged to villages, large villages, or cities. Minimally, such indicators as population density within settlements should be included into the classification typology; this would give an objective idea of the compactness or building or rareness of domestic development in percentage terms. To determine the factor of the settlement functional focus and the principles of the object space-planning pattern, taking into account the state of its preservation is of the essence as well.

Based upon the dwellings planning on the fortified settlements, they were intended for congeneric representatives. Indoors, the dwellings were divided into small rooms intended for a small family of 4–5 people. The average number of inhabitants in one dwelling is calculated as per the formula listed above:

$$(\mathrm{Xmax} + \mathrm{Xmin}) / 2 = (40 + 30) / 2 = 35\,\text{people}, \quad (1)$$

where Xmax is the maximum number of inhabitants in one dwelling of the Sintashta fortified settlement (according to archaeology [15]);

Xmin is the minimum number of inhabitants in one dwelling of the Sintashta fortified settlement (according to archaeology [15]).

Thus, we obtain the average gross floor area per capita:

$$S(overall) = 115 / 35 = 3{,}3 sq.m / capita. \quad (2)$$

$$S(overall) = 90 + 140 / 2 = 115\, sq.m; \quad (3)$$

The average area per family of four people:

3,3 × 4 = 13,2 sq.m.

According to modern concepts, these figures, if archeologists' studies are to be trusted, seem extremely minimal.

Currently, it is not anticipated to introduce a population density criterion, since the current state of the issue does not have sufficient data on population density in the territories of the Southern Urals fortified settlements during their functioning. The data on the proposed number of inhabitants in various sources vary greatly and require further study before these indicators will be included in the typological assessment criteria.

According to the V.F. Gening study materials ([15], p. 24), it is necessary to rest upon the calculations of the Sintashta-1 settlement area, taking into account the adjacent territory, which reaches at least 6.2 hectares, therefore falling under the large object subtypes. Thus, the Sintashta-1 fortified settlement may be considered the base for determining the standard size criterion.

All fortified settlements that are smaller in terms of area than Sintashta-1 fall under the 1st subtype; some of these two subtypes have a two-level system of fortification, that is, consist of two rings of defensive works and dwellings: internal and external. Some do not have an internal defense ring (they either lost it in the process of rebuilding and functioning or did not have it initially); such constructions have a fairly vast courtyard space.

Upon closer inspection of the Sintashta fortified settlements of two-level fortification, such as Arkaim, Sintashta-1 Kizilskoye, possibly Bersuat or the early layers of Steppe, Kuisak, based on the correlation of archeological data decryption of aerial photographs by B.V. Adrianov and I.M. Batanina [16], are considered as fortified settlements, which initially have had a "citadel" in the structure of fortification, around which an outer ring of inhabitable walls was formed in stages, forming the second level of fortification.

In the course of the evolution of scientific study and fortified settlements' ruin deciphering, such as Sarym-Sakly and Aland, it was found that initially, these settlements did not have an internal citadel at all, or lost it in the process of reconstruction. If such a structure previously took place on the settlements site, then it functioned as an earlier form of construction, on the basis of which, later, a larger settlement arose, and the citadel itself was lost in the process of reconstruction.

On the example of the Andreevsky fortified settlement, the plan shape of the Sintashta type has not always been round or oval, such as in Arkaim, Sintashta, Sarym-Sakly, and others. The plan shape of the Sintashta settlements, presumably, depended on the configuration and shape of the ground. By the graphical reconstruction of vertical planning pattern and the construction of horizontal sectional view of the plan, it can be inferred that all round-planned settlements are located on a table land and radically have the shape of a "saucer". Thus, this shape was associated not only with defensive but also with the hydrotechnical functions of the construction, protecting it from the force of nature.

Since the mid-1980s, the "landscape archaeology" dimension has been engaged in the study of the processes of settlements' adaptability and resistance to environmental conditions in different periods of the settlement systems development, from the Paleolithic to the ancient period. "...serious progress is provided not by new epigraphic significant sites rare finds, but by the vast amount of data on the artifacts under study. This knowledge deepening progress can be provided by a complex of methods for studying the interaction of humans and the surrounding landscape" ([17], p. 143–156).

During the floodtime, the site with the settlement provided an island location of the settlement, surrounded by flood waters. Not only the settlement itself remained on land within the boundaries of the defense construction, but the adjacent territory area was also not exposed to flooding.

As a result of the Andreevskoye fortified settlement analysis, it can also be concluded that the plan shape of this settlement is due to the landscape shape: the settlement was inscribed in the riverbed terrace upland.

Initially, by shape, Andreevskoye looked like a citadel with distorted rectangle and rounded corners. Excavations in the Andreevskoye settlement have not yet been carried out; however, in 2013, major geomagnetic and geo-radar studies were carried out by V.V. Noskevich and N.V. Fedorova, the researchers of the Institute of Geophysics of the Ural Branch of the Russian Academy of Sciences. As a result, it was found that the significant sites' ruins had the form of overlays of three asynchronously fortified settlements with a shape similar to a rectangle and the earliest of which was synchronous with the Sintashta period. Furthermore, it was found that the construction was built about 4000 years ago, in the form of a rectangle, and was subsequently finished and rebuilt several times [18].

The issues of chronological component in the development of the fortified settlements typology are not touched upon in this study advisedly, in order to avoid affecting the objectivity of the formation assessment due to disagreements in the cultural layers dating. According to the same approach, Walter Kristaller, the German scientist-geographer, studied settlement systems regardless of their historical background, and this allowed achieving certain results in a universal model development ("W. Kristaller's model of the hierarchical structure of central places") [19]. Later, the model has been used and supplemented in the theory of urban planning as one of the basic ones. It has also applied to ancient Mesopotamian settlements [20] where stable indexes of distances between settlements of different ranks were traced, which was inextricably linked to the transportation facilities of the population of that period.

In the rectangular fortified settlements, unlike round-planned settlements, there was no central element—a land area, and there were also no hydraulic ring ditches. Drain was carried out toward the relief; perhaps other drainage canals were laid along the walls and dwellings. Similarly, later settlements of the Petrovka or Srubno-Alakul type were built, characterized by a kind of a continuous building pattern. For such site development, this is the most sustainable solution—the creation of longitudinally axial streets inside the settlement and the building of consolidated dwellings isolated from the fortification. This technique made it possible to rebuild and finish constructing residential and fortification structures and change the configuration of the settlement without affecting residential areas. The fact of such frequent completion and reconstruction of rectangular-type settlements is evidenced by their multi-layered nature.

The Andreevskoye fortified settlement is an extremely important object for understanding the transformation of the typology, since it is this settlement that can be a borderline in the transformation from the Sintashta to Alakul (Petrovka) type of plan.

Andreevskoye demonstrates a transition force from the Sintashta cultural deposit to the Petrovka (Srubno-Alakulsky) [21]. According to the type of building, it is at the initial construction phase with inhabitable walls, but according to the plan form, it is characteristic with fortified settlements with a continuous building pattern.

Such fortified settlements as Rodniki, Steppe, Olgino (Kamennyj Ambar) [22] possibly Kuisak, Kamysty, and and Ustye were multilayered and had all the cultural layers of the Bronze Age. Thus, during steps of the determination of the morphotype, special attention should be paid to the configuration of the initial layer planning structure.

The following that should be focused in the development of the fortified settlements typology on is the fortification system. When analyzing the vertical planning, three different types of fortification have been identified.

In the fortification strength assessment, the term "areotectonics" has been used since the XVIII century. ("Areotectonics is a complex of processes for the of defense works construction" ([23], p. 25).

For convenience of estimating the Sintashta-Petrovka areotectonics characteristics, an assessment level of its complexity or strength is introduced, which is determined by 3 indicators: depth (horizontal), height (vertical), and the number of barrier levels (walls, ramparts, and ditches in the fortification depth).

The 1st degree of fortification complexity, the Sintashta type (Illustration 1A) – this type may be found only around a free-standing "citadel" or in large fortified settlements that have the form of structures without an internal citadel (perhaps there was a citadel early in settlement development).

The peculiarity of this type was that the fortress had two parallel defensive walls, between which there was a drainage ditch. The outer wall in a circle or perimeter had the form of a rampart pattern with indoor premises up to 2.5–3 meters high and up to 3 meters wide at the base, and the inner wall, to which the dwellings adjoined, looked like a complex fortification, and communication system with the width reached 6.5 meters at the base and up to 4 meters at the top of the wall. The parapet height reached 5.5 m sub-mainland. In some sources, this wall is called the "upper circular street" [5]. From the outside, a quite deep drainage ditch up to 3 meters deep was adjacent to the wall. There were premises-cells inside the wall, in which it was possible to get from the dwellings adjacent to the fortification. In some cases, the walls were faced with stone slabs and fieldstones (granites, amphibolites) from the outside. The depth of such fortification could reach 16 meters, and the total height could reach 8.5 meters, taking into consideration the drainage ditch.

Despite the fact that such fortification is specific to large Sintashta fortified settlements, such as Aland and Zhurumbai, it is also found in small "citadels": Sarym-Sakly or Kuisak (during the opening phase of formation) as an example, possibly Isinei (excavations at the settlement have not been carried out).

The 2nd degree of fortification complexity, the Sintashta type (Illustration 1B) - having completely and thoroughly studied Arkaim, Sintashta, and other fortified settlements, a description of this fortification as "Sintashta" is given. The second level of complexity may be assigned to the external defensive walls of the Arkaim, Bersuat, Sintashta, Kizilskoye, and other fortified settlements. This type of fortification and communication system consisted of a defensive wall, to which the dwellings adjoined. Behind the wall, there was a small scarcement, to which an areaway adjoined. A dirt ridge with a earthen rampart was formed behind the areaway. The width of the rampart reached 5–6 meters at the base and up to 3 meters at the top of the wall; the parapet height reached 2.5–3 meters sub-mainland. There was a manhole or exit to the wall from each dwelling. Thus, the depth of such fortification system was 9–12 meters, and the total height was 5.5–6 meters, taking into account the bypass ditch depth (**Figure 1**).

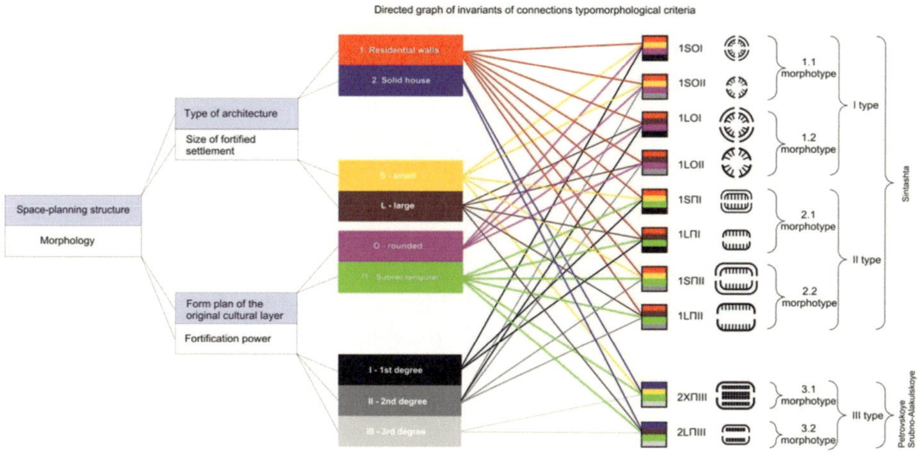

Figure 1.
Sintashta-Petrovka areotectonics (graphic reconstruction). (A) the 1st degree of fortification complexity. (B) the 2nd degree of fortification complexity. (C) the 3rd degree of fortification complexity.

The 3rd degree of fortification complexity, the Srubno-Alakul type (Illustration 1C) - the Olgino (Kamenny Ambar) and Ustye fortified settlements have been virtually reconstructed and studied to the fullest extent possible. The defensive system of this type was a wood-soil rampart in the form of a wall up to 2–2.5 meters high on the outside faced with stone slabs or raw blocks, bonded with clay mortar, surrounded by small ditches or drainage ditches on both sides. Behind the outer ditch, an external clay embankment or a dirt ridge was poured, to which an external ditch or drainage ditch adjoined.

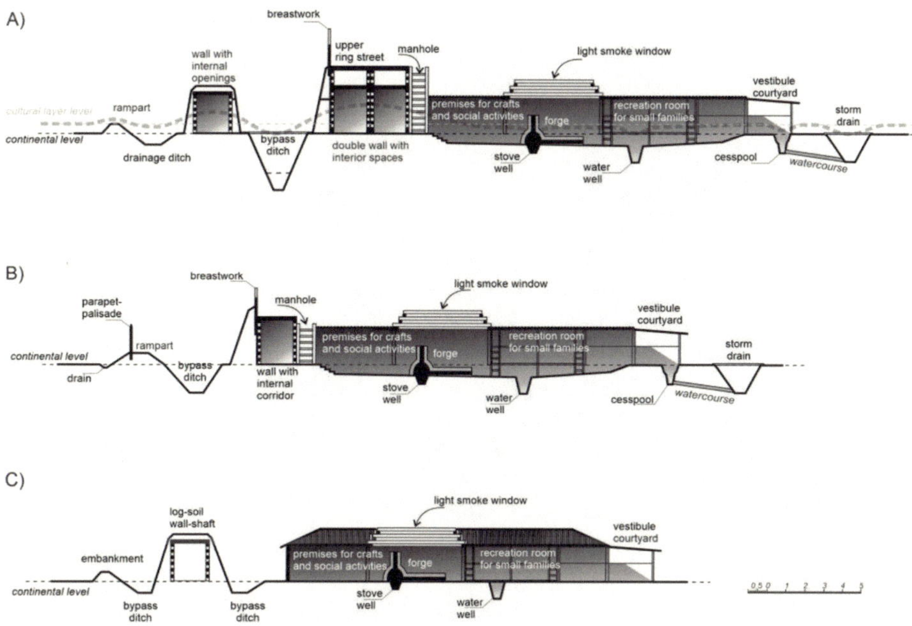

Figure 2.
Typology of fortified settlements of the bronze age in the southern Urals.

Due to the fact that excavations and surface magnetic studies have not been carried out on all Southern Urals fortified settlements, it is not possible to determine the type (level of complexity) of fortification on most settlements at the time of the study.

Based on the analysis, it is possible to identify the main criteria by which it is possible to determine the Southern Urals of the Bronze Age fortified settlement typology.

As a result of the methodology application, based on the definition of typological criteria for assessments, a typological scheme of Southern Urals of the Bronze Age fortified settlements has been formed. Two key indicators, which are determined by the generalized characteristics of the study objects, space-planning pattern and morphology [24], form the basis for criteria for assessments and typological features.

The space-planning pattern is estimated by the architecture type and the type of planning pattern, with inhabitable walls/with a continuous building pattern, and by the fortified settlement size, small/large.

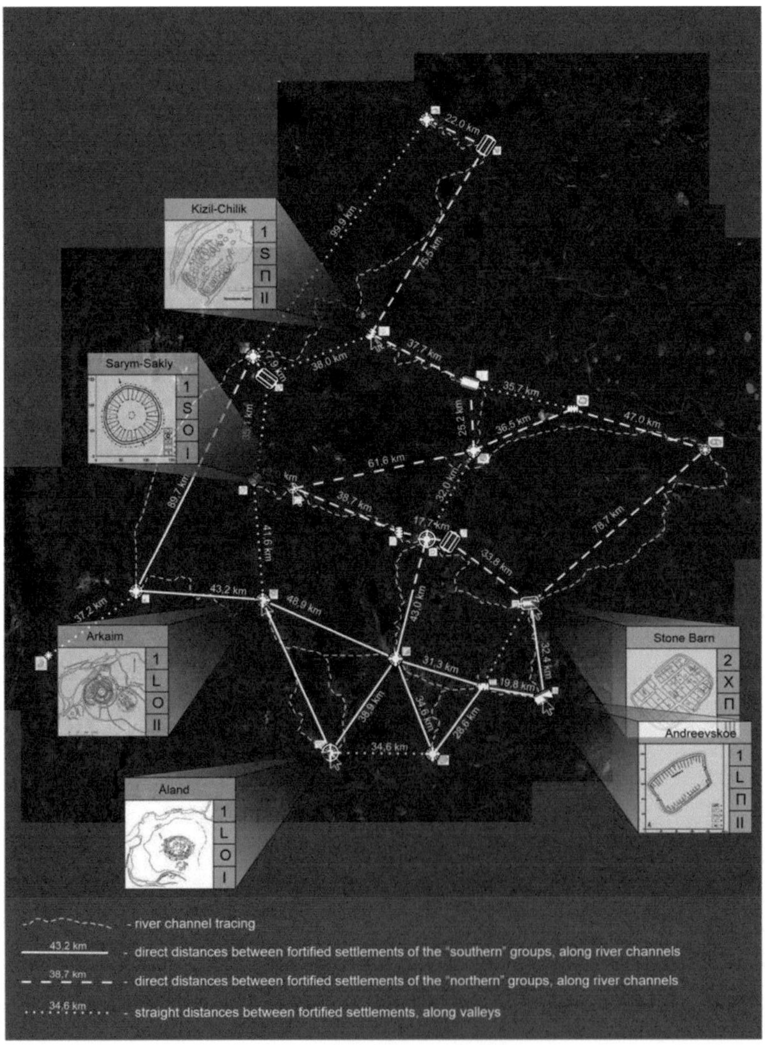

Figure 3.
Approbation of the typological scheme application results in practice.

The fortified settlement morphology is estimated by the plan shape of the initial layer, rounded/subrectangular, and by the fortification strength.

As a result, a graphical scheme has been drawn, which represents the main typological classification of the Sintashta and Petrovka (Srubno-Alakulsky) period fortified settlements in the Southern Urals (**Figure 2**) at the time of the study.

The developed scheme's advantage is that it is dynamic. Only the main (basic) parameters remain unchanged in it, and criteria may be added or updated in the case of data correlation or the identification of new typological criteria.

In evaluation of study results and as an example of practical implementation, an analysis of the objects location on a satellite map using the developed typology is given (**Figure 3**). The diagram demonstrates information on anchor points (anchor points indicate fortified settlements). Pictograms on the diagram indicate fortified settlements by morphotypes; linear distances between anchor objects and anchor features are presented (displayed with lines of different types). When mouse over to any of the anchor points, a "scroll" with information appears. The interactive block in the "scroll" contains all the necessary information about the anchor point: name, image, and basic typological parameters.

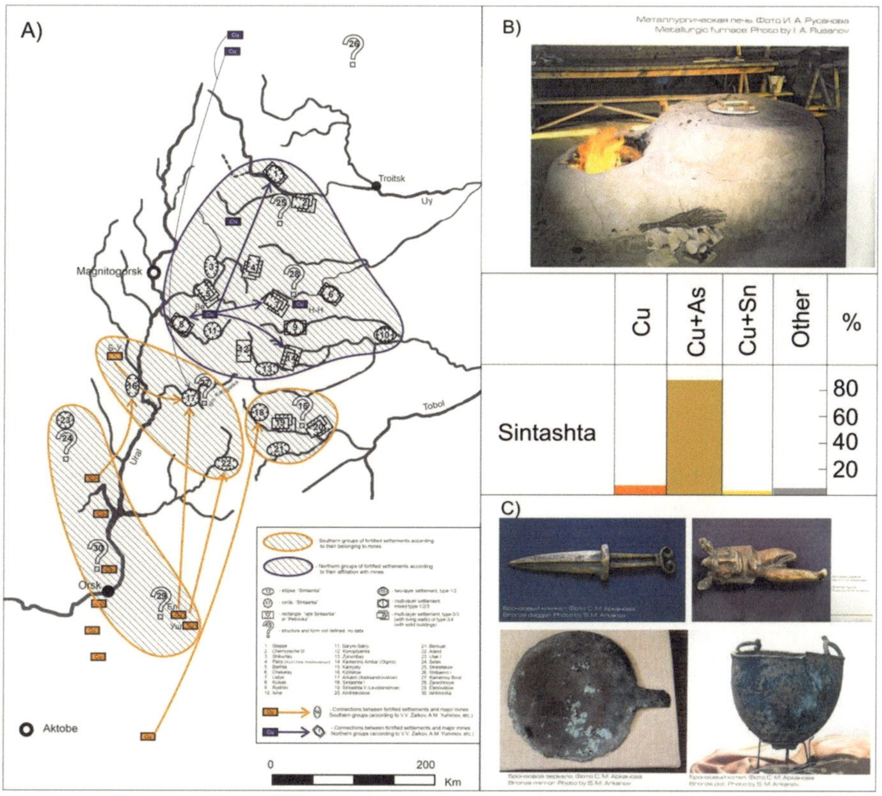

Figure 4.
Metallurgical production in the southern Urals territory in the bronze age with the analysis of the resource potential and different types of fortified settlements belonging to the raw materials extraction sites. (A) O.A. Ulchitsky, 2021. Fortified settlements groups in relation to resource potential. (B) F.N. Petrov, 2008. Reconstruction of the Sintashta iron-and-steel furnaces [25] and the alloy types ratio on Sintashta monuments. (C) F.N. Petrov, 2008. Objects made of bronze in the Sintashta settlements [25].

2. Sintashta metallurgy and bronze

Briefly, it is worth noting that the Sintashta bronze metallurgy belongs to one of the latest stages of the Middle Bronze Age. At that time, bronze manufacturing technology demonstrated a sufficiently large inventory assortment from housekeeping equipment and votive items to weapons (**Figure 4**), which, obviously, as trophies were a great asset to their owners.

On the fortified settlements territory, objects made of metallurgical alloys, the main composition of which was copper and antimony, have virtually been not found (there are isolated finds), but inventory of its kind was mainly part of the Kurgan tombs of the Bronze Age and later period. The main and most interesting finds were made on the territory of the Arkaim and Sintashta burial grounds. Smelter slag was found in large quantities on and around the settlements, as well as spatters and molds [26]. This topic is dug deeper into the I. V. Chechushkov dissertation [27].

3. Conclusion

The study extension on the development of the fortified settlements' typology in the Southern Urals at the end of the III to the first half of the II millennium BC provides a more accurate and overall picture of the evolution formation of ancient settlements.

The new developed typology of ancient fortified settlements of the Southern Urals essentially fills up the previously known developments. It specifies and classifies objects much more precisely according to the identified criteria. Generally, the developed typology specifies 3 main types of the Southern Urals fortified settlements of the Bronze Age, (1) Sintashta round-shaped "with an inhabitable walls pattern," (2) Sintashta subrectangular "with an inhabitable walls pattern," and (3) Petrovka or Srubno-Alakulsky subrectangular "with a continuous building pattern, which are combined into six morphotypes.

The study continuation on the development of the fortified settlements typology in the Southern Urals at the end of the III to the first half of the II millennium BC provides a more accurate and overall picture of the settlements' oldest forms evolution, which represents a linear, cluster, and poly- or monocentric settlement system over a vast territory.

The developed typology forms ten subtypes, which are combined into six morphotypes.

The developed scheme is a good example of typological connection invariants possible in theory and may be used as a tutorial for studying fortified settlements ruins on topographic maps at various scales and in the construction of linear structures conjoint by typological connections.

The theoretical significance of the study results' approbation lies in the fact that the graphical display of information about objects and their connections provides operational possibilities for visualizing data on anchor points (fortified settlements) on satellite maps. The practical significance of the implementation lies in the prospects of creating a software module pegged to GIS systems.

DOI: http://dx.doi.org/10.5772/intechopen.1004078

Author details

Ulchitsky Oleg Alexandrovich
Department of Architecture and Fine Arts, Nosov Magnitogorsk State Technical University, Russia

*Address all correspondence to: ulchitsky@gmail.com

References

[1] Gurevich LL. O kruglyh gorodah drevnih irancev. In: Zony i etapy urbanizacii: Teoreticheskie aspekty problemy Gorod i process urbanizacii v Srednej Azii Tez. dokl. konf. Namangan. Tashkent. 1989. pp. 48-49

[2] Petrov FN, Ankushev MN, Medvedeva PS. Material'nye svidetel'stva tekhnologicheskih processov v kul'turnom sloe poseleniya Levoberezhnoe (Sintashta II): opyt funkcional'nogo podhoda, Magistra Vitae. elektronnyj zhurnal po istoricheskim naukam i arheologii. 2018;**1**:112-147

[3] Chechushkov IV, Dakovich G, Yakimov AS. Arhitekturno-planirovochnye resheniya sintashtinsko-petrovskih poselenij i problema voennogo dela epohi bronzy. Rossijskaya arheologiya. 2018;**2**:75-93

[4] Zdanovich GB, Batanina IM. Arkaim, Strana gorodov: Prostranstvo i obrazy. Chelyabinsk: Krokus; 2007. p. 260

[5] Malyutina TS, Zdanovich GB. Alandskoe–Arkaim: drevnee nasledie Yuzhnogo Urala. Orenburg. Izd-vo Ministerstva kul'tury, obshchestvennyh i vneshnih svyazej Orenburgskoj oblasti {Orenburg: Publishing House of the Ministry of Culture, Public and External Relations of the Orenburg Region. 2013. 30 p

[6] Soldatkin NV. Arhitektura ukreplennyh poselenij sintashtinsko-petrovskogo tipa v sushchestvuyushchih variantah vizual'nyh rekonstrukcij. Vestnik arheologii, antropologii i etnografii. 2020;**1**(48):15-25

[7] Ul'chickij OA. Osobennosti arhitektury protogorodov drevnego Urala (na primere Arkaima): avtoref. Russia: dis. ... kand. arhit. Ekaterinburg; 2006. 24 p

[8] Masson VM. Dinamika razvitiya tripol'skogo obshchestva v svete paleodemograficheskih ocenok. In: Pervobytnaya arheologiya: poiski i nahodki. Kiev: Naukova dumka; 1980. pp. 204-212

[9] Fyodorova NV, Noskevich VV. Rekonstrukciya planirovki ukreplennyh poselenij epohi bronzy na YUzhnom Urale (Ol'gino i Konoplyanka) po rezul'tatam detal'noj magnitnoj semki. Ural'skij geofizicheskij vestnik. 2012;**1**:52-59

[10] Fyodorova NV. Geofizicheskie metody issledovaniya arheologicheskih pamyatnikov Sarym-Sakly i Vorovskaya Yama (Yuzhnyj Ural). Ural'skij geofizicheskij vestnik. 2013;**2**(22):46-53

[11] Ulchitsky OA, Bulatova EK, Kazaneva EK, Veremey OM. A comparative study of the layout of bronze age fortified settlements in the southern Urals (3rd to 1st millennia BC). Archaeology, Ethnology & Anthropology of Eurasia. 2019;**47**(1):64-72

[12] Ponomarenko EV. Goroda Yuzhnogo Urala. Vol. Tom I. Chelyabinsk: Izd-vo Chelyabinskogo oblastnogo komiteta statistiki; 2005. p. 24

[13] Tolstov SP. Po sledam drevnekhorezmijskoj civilizacii. M.; L.: Izd-vo AN USSR; Moscow-Leningrad: USSR Academy of Sciences; 1948. 328 p

[14] McC AR. Land behind Baghlad. A History of Settlements on Diala Plants. Chicago, London: University of Chicago Press; 1972

[15] Gening VF, Zdanovich GB, Gening VV. CH. 1. In: Sintashta: Arheologicheskie pamyatniki arijskih plemen Uralo-Kazahstanskih stepej: v 2 ch. Chelyabinsk: Yuzh.-Ural. kn. izd-vo; 1992. 408 p

[16] Batanina NS, BRK H. Chapter 12. Soviet period air photography and archaeology of the bronze age in the southern Urals of Russia. In: Hanson WS, Oltean IA, editors. Archaeology from Historical Aerial and Satellite Archives. Springer Science+Business Media, LLC; 2013. pp. 199-219. Chapter 12. DOI: 10.1007/978-1-4614-4505-0_12

[17] Malyshev AA, Trebeleva GV. Evolyuciya sistemy rasseleniya v periferijnyh rajonah Aziatskogo Bospora (na primere p-va Abrau). In: Problemy istorii, filologii, kul'tury. Vol. 3. 2018. pp. 143-156. DOI: 10.18503/1992-0431-2018-3-61-143-156

[18] Noskevich VV, Fedorova NV, Bebnev AS, Vdovin AG, Mekhonoshina TL. Rezul'taty issledovaniya geofizicheskimi metodami arheologicheskogo pamyatnika bronzovogo veka gorodishche Andreevskoe (Yuzhnyj Ural). Ural'skij geofizicheskij vestnik. 2014;**1**(23):72-80

[19] Christaller W. Die zentralen Orte in Süddeutschland. In: Eine ökonomisch-geographische Untersuchung über die Gesetzmäßigkeit der Verbreitung und Entwicklung der Siedlungen mit städtischer Funktion. Darmstadt: Wissenschaftliche Buchgesellschaft; 1980. (Repr. d. Ausg. Jena 1933)

[20] Adams R, Nissen HJ. The Uruk countryside. Chicago-London: University of Chicago Press; 1972. 241 p

[21] Epimakhov AV. Multidisciplinary archaeological research in the study of migration and mobility (bronze age of the Southern Urals). Problemy istorii, fi lologii, kul'tury. 2021. p. 5-17. DOI: 10.18503/1992-0431-2021-3-73-5-17

[22] Koryakova LN, Krause R, Epimakhov AV, Sharapova SV, Panteleyeva SE, Berseneva NA, et al. Archaeology Ethnology & Anthropology of Eurasia. 2011;**39**(4):61-74

[23] Artyuhovich DV. Areotektonika, Stroitel'stvo. In: Enciklopedicheskij slovar. Stavropol: Paragraf; 2011. 766 p

[24] Yankovskaya Yu. S. Obraz i morfologiya arhitekturnogo ob'ekta / YU. S. YAnkovskaya. Ekaterinburg: Arhitekton; 2004. 91 p

[25] Nikolaevich PF. Arkaim i Strana gorodov: istoriya i priroda stepnogo Zaural'ya: [fotoal'bom/avt. teksta i sost]. Chelyabinsk: Kprkus; 2008. 72 p

[26] Prihod'ko VE, Ivanov IV, Zdanovich DG, Zdanovich GB, Manahov DV, Arkaim IK. ukreplennoe poselenie epohi bronzy stepnogo Zaural'ya: pochvenno-arheologicheskie issledovaniya. M.: FGUP Izdatel'skij dom "Tipografiya" Rossel'hozakademii. Federal State Unitary Enterprise; Typography, Russian Agricultural Academy Moscow. 2014. 264 p

[27] Chechushkov IV. Bronze Age Human Communities in the Southern Urals Steppe: Sintashta-Petrovka Social and Subsistence Organization [Thesis]. Pittsburgh: University of Pittsburg; 2018. 214 p

Chapter 2

Copper and Flint – Exploring Technological Interfaces in South Scandinavian Early Metal Using Societies

Moiken Hinrichs, Khurram Saleem, Berit V. Eriksen and Lorenz Kienle

Abstract

This chapter explores research pertaining to the possible evidence of copper tools being used in flint knapping processes in early metal-using societies in South Scandinavia. The existence and use of copper as a flint working tool in prehistoric Scandinavian contexts has often been proposed and accepted, but no real study on the implementation as such or the effects on the knapping process has been conducted. In the absence of archaeological evidence, the proof and interpretation must rely on secondary markers, such as technical attributes or trace elements on the flint. Research so far has analysed finished tools to detect and verify the use of copper. This study offers a different approach, relying on production flakes to get a deeper understanding of the production process itself and thus investigating the effects copper knapping tools would have had.

Keywords: prehistory, metallurgy, processing, trace analysis, technological transformation

1. Introduction

In flint knapping, the oldest known and most persistent technique is direct percussion with a hard hammer stone. First, evidence has been traced back 3.3 million years, long before the existence of the genus *Homo*, e.g. [1, 2]. For the longest time, it has been the only way of working stone tools. Only some 100,000 years ago, during the late Lower Palaeolithic, a change in manufacturing methods appeared, and organic percussors enter the knapping process [3, 4]. By comparison, the discovery and processing of metals happened quite recently, only 5000 to 6000 years ago, during the Chalcolithic, after which it started to spread at varying speeds through Europe [5, 6].

Invention and adoption of metallurgy most certainly was one of the most powerful technological milestones in prehistory [7, 8]. Not only did the objects made from a novel material have an influence on the societies, but also the knowledge and possibilities tied to the invention acted on people as well as on the existing technological traditions. Yet, no technology appears out of thin air; ideas, procedures, and developments are often based on existing knowledge adapted to new circumstances. This is also the case concerning metallurgy. The material is new, and the properties are very different, but the idea of heating material in order to change its properties is not a novelty. This process has a longer tradition in ceramic production and even flint knapping [7, 9]. The challenge in the processing of the new material was to achieve higher and steady temperatures so the ore could smelt [10–12]. Metallurgic knowledge and the tools created were thus not separate entities but part of the entire technological system. They were influenced by and did influence other technological aspects, like flint knapping.

While questions of technological developments in local metallurgy and its effects on the social sphere have been analysed, e.g. [13–15], the effects on existing local technologies and traditions have only received little attention and then mostly pertaining to the decline of the flint knapping craft, e.g. [16–18]. This article examines the influence of different technological complexes, here, flint working and early metallurgy, on each other and explores the possibilities for identification in archaeological contexts.

2. Research background

The emergence of civilisation and social hierarchisation has often been treated as an inevitable development connected to the invention and spread of metallurgy. The new prestigious material as well as the new developing technology, implied new social structures. On one hand, the complex extraction and/or trade and processing of metals needed specific working structures for a successful outcome and on the other hand, this eased the path to individuals seizing control over raw material access, craft knowledge, (trade) routes, and other influential and necessary aspects of life. This can be sensed in changing burial rites, including a shift from communal megalithic burial to family or individual-oriented mounds, but also in the changing symbolism and objects connected with certain social rules and genders as well as settlement patterns and house construction [15, 19]. However, the adoption of early metallurgy has not been a straightforward path in the development of societies [9]. Northern Europe, and especially Southern Scandinavia, is a good example of a culture-geographic area where an initial familiarisation with the novel material was not immediately followed by an adoption of the new technology. The earliest copper objects in Scandinavia have been found in Late Ertebølle (Mesolithic) contexts and date around 4500 BCE [14, 20]. Based on typological similarities and metallographic analyses, the pieces have been identified as imports from Southeast Europe [14]. Early objects are typically tools like axes, but awls, beads, and rings are also known [14, 20]. Large-scale adoption of metallurgy did not happen; likewise, local copper sources, especially in today's north Sweden, were not exploited in prehistoric contexts, although mining was a familiar practice, as flint quarries indicate [21–24]. All copper had to be imported to Scandinavia, either as finished tools or in the form of ingots [16, 25, 26]. New data suggest that some experimentation with metallurgy in the early phase in the form of cold forging and probably also smelting was present [6, 27, 28]. Likewise, knowledge about physical properties is hinted at in prehistoric finds [29, 30]. After an initial boost, copper objects disappeared from archaeological contexts between 3200 and 2350 BCE [6, 9], which is often interpreted as an indication of changing or collapsing networks, leaving Scandinavia at the fringes of the technological development without possibilities to participate, e.g. [9, 31]. Opposing opinions and explanations have been voiced. Missing evidence does not indicate absence *per se*. Changing depositional practices or intensified re-melting are likewise possible explanations for the absence of objects [6, 28]. However, metallurgical analyses have detected changing isotope compositions in the copper alloy of the objects from the initial phase and objects from the Late Neolithic and Bronze Age context, which favours the interpretation of changing networks, also indicating that metal is imported from different regions than initially after the hiatus [6, 9, 32]. Likewise, throughout the Bronze Age, certain types and forms of objects and materials appearing in different European regions indicate a developing and changing long-distance trade network connected to the distribution of metal across Europe, where Scandinavia contributed by supplying central Europe and the Mediterranean with amber [15, 19, 33, 34].

Another option for the decline is that copper tools simply were not as much in demand as always assumed. The early copper tools especially did not present technical advantages to the existing tools made of bone, stone, or flint, as the copper was far too soft to offer the same efficient use [35, 36]. Early copper objects in Scandinavia are often found in hoards and in pristine condition, which suggests a purely symbolic function. It is the aesthetics and rarity of the material which made it desirable, not the technical function [37]. The notion of metal as a superior and prestigious material is a modern view of prehistoric times, which needs to explain the adoption or abandonment of metallurgy [9, 36, 38]. However, copper has technical advantages when used in flint knapping. In comparison to antler or wood, copper is a harder yet still flexible material. Antler pressure flakes need a certain size as the contact area of the tip during the flaking process, as the applied force otherwise will cause the tip to splinter (**Figure 1**). The physical properties of copper extend the possible force that can be applied during the flaking while reducing the contact area of the tip; the material will deform a bit but will not splinter. In other words, as the material is harder than organic materials, the tip can be smaller and more pointed, and as it is still flexible, the tip will withstand the pressure without shattering. Likewise, the reduced contact

Figure 1.
Flint knapping tool kit by PW in 2007. Photo by: B.V. Eriksen.

area provides a more concentrated transmission of force, which relates to less force being needed from the knapper's side to remove flakes. On the downside, as the copper tip is harder, it is more likely to result in a conchoidal fracture, which often leaves deeper negatives than pressure with an organic flaker would. This means additional work has to be invested to achieve an even working surface.

The need to import metal objects and the shortage of the very same material has been used as an explanation for the development of bifacially[1] crafted flint daggers in Scandinavia between 2350 and 1600 BCE. It also has been frequently stated that the daggers are skeumorphic alternatives to copper daggers made from locally available material [39–44]. But the bifacial flint knapping method[2] is older, and other types of bifaces were manufactured in Scandinavia before the advent of daggers [46–48]. Similarly, the typological development of the daggers appears continuous without breaks or jumps in the overall shapes. Metal objects seem to have had a guiding effect on the forms of flint tools but were not necessarily the origin.

Scandinavian flint daggers were produced and circulated in northern Europe for about 800 years. Throughout this time, a variety of forms developed, which were divided into six types based on the form of the blade and handle by E. Lomborg [49]. Each type consists of at least two subtypes (**Figure 2**), but from a morphological point of view, the daggers can be summarised into four categories: lanceolate (types I and II), lanceolate blades with square-profiled handles (type III), leaf-shaped blades with so-called fishtail handles (types IV and V) and leaf-shaped blades with tang-like handles (type VI). They are chronologically more or less successive, although with larger overlaps and regional variations in type preferences [16, 49, 51]. Especially the so-called fishtail daggers (**Figure 3**) have received considerable attention due to their impressive shapes, difficult and complex manufacturing process, and likeness to contemporary bronze daggers in northern and central Europe [42, 52]. Without a doubt, Late Neolithic and Early Bronze Age flint knappers were experts in the craft and pushed the boundaries of the physically possible. A similar perfection and command of the flint knapping process has not been observed before or after in Europe.

Starting with the type III daggers, punched seams are created on mostly rhombic handles, which are often interpreted as copies of metal dagger seams created by the casting mould [53, 54]. It is these handles that have puzzled generations of archaeologists and modern flint knappers. Fracture mechanics specify certain angles at which successful removal of flakes can happen. Ideally, depending on the technique used during the reduction, an angle between 60 and 80° is aimed for [45, 55]. The seams on the rhombic handles are often worked at a 90° angle on all four sides. Furthermore, the knapper had to change from planar to four-sided working mode on the same piece. Although the technique employed to work those forms was long incomprehensible from a modern point of view, studies have shown that the knowledge and working tradition was nothing new for flint knappers in the Late Neolithic and Early Bronze Age. The method has a long tradition in the manufacture of the four-sided axes and adzes of flint from the region [56]. Creating 90° seams on flint objects is thus nothing really new in South Scandinavian flint knapping traditions, but the filigree character of the stitching, as seen in **Figure 3**, represents a refinement of existing knowledge and methods. Quite likely related to the possibilities copper offered in flint knapping.

---

[1] As the term indicates, bifaces are flint objects, which have been worked on both sides of the nodule – across both faces of the piece.

[2] The article follows the French terminology. Method is thus defined as a sequence of interrelated actions, which lead to the production of predetermined products [45].

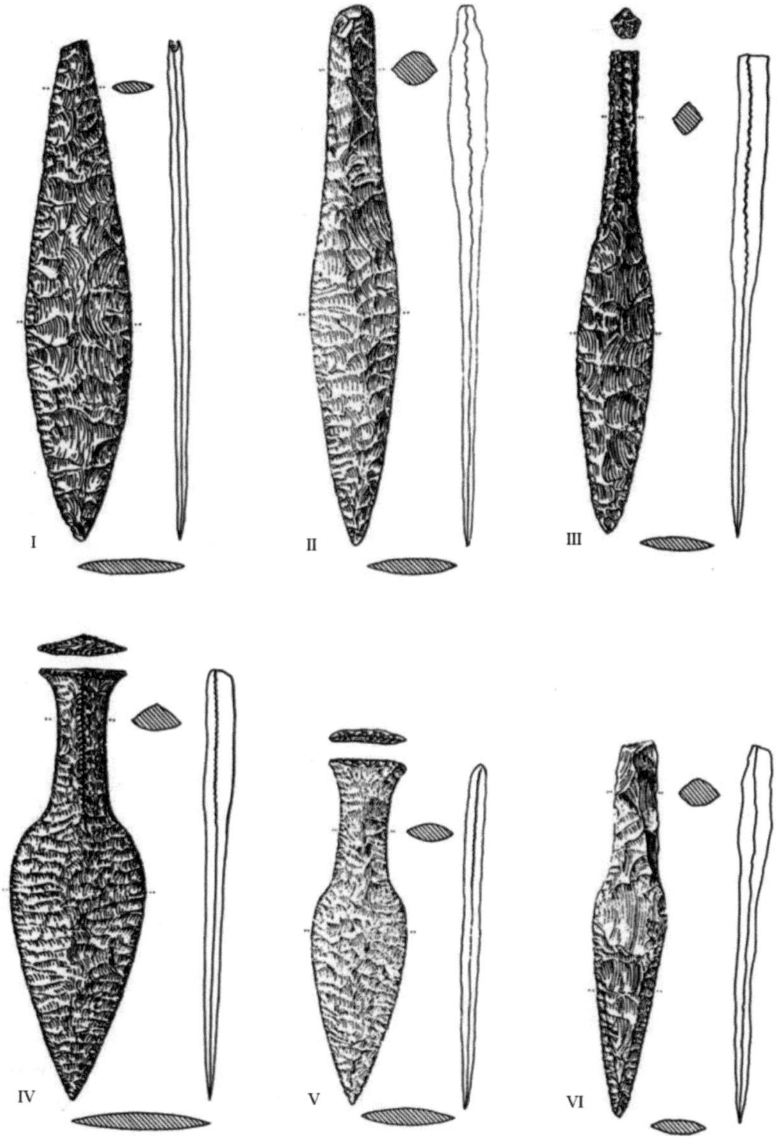

Figure 2.
Overview of flint dagger types based on E. Lomborg's [49] drawings assembled by J. Apel [50].

It is unclear whether the manufacture of the handle was done like this for aesthetic or, indeed, for technical reasons. Flint is a hard and, in flaked form, very sharp material. It is quite uncomfortable to wield a freshly made dagger in bare hands, and it can result in deep cuts if not treated carefully. The lanceolate daggers were likely used hafted (i.e., with a wrapping of the handle made of organic material), as finds from different regions suggest [57], for example, the dagger from Wiepenkaten [58] and a recent find from the Femern belt excavation [59–61]. The stitched seams are still mostly interpreted as an aesthetic attribute, which was meant to be shown, especially on the type IV fishtail daggers. Points in favour of this interpretation are that the handles of the earlier dagger types are often not worked as carefully as the blade, probably due to the later wrapping, which would cover the surface. Likewise, the

Figure 3.
Type IV flintdagger. Holstein-Rathlou collection, Moesgaard Museum, Denmark. © Rógvi N. Johansen, Moesgaard Museum.

stitched seams often show grinding to blunt the sharp ridges, which would lessen the danger of injuries during handling. Another explanation for the seams is the possible advantages of the wrapping of the handle. The organic material would have a better grip and be less prone to slip on the seams. An example in favour of this interpretation is a reported type V dagger, which allegedly was found with a thread wrapping which, unfortunately, was removed by the collector [48, 62].

The manufacturing process of the flint daggers is a complex series of interrelated steps. Depending on the type of dagger, up to eight stages may have to be completed before a finished tool is produced [63]. Simply put, the more elaborate the dagger was meant to be, the more stages and more careful work were needed. A point which has often been stressed is that the development of flint crafting was only made possible due to the rise of metallurgy and the inflow of copper daggers into Scandinavia. The scarcity of metal and the determination of lithic craftspeople not to be left behind started a technological race which prompted the perfection of the flint knapping process and resulted in more and more elaborate and stunning dagger designs [40, 64]. Moreover, there are strong arguments to suggest that the development also did benefit directly from the upcoming metallurgy. Certain steps in the manufacture of some dagger types are not possible or, at any rate, very difficult to master without the use of copper-tipped implements, i.e., the so-called pressure flakers. This concerns the finishing parallel retouch of the blade surface of the type IC daggers but also the finely stitched seams of the type IV daggers [53, 63, 65, 66]. The necessity of including copper-tipped pressure flakes to achieve fine parallel flaking is still debated in the flint knapping community. While all agree that some of the finest seams of the fishtail daggers show too small negatives of removed flakes to be done by anything other than copper tips, the parallel flaking is more debatable. Some flint knappers suggest that it could be possible to achieve similar results while using spatula-shaped antler flakers [67], which do exist in the archaeological record and could be related to pressure flaking [68]. Regardless of the possibility of using different materials and tools for the parallel retouch, copper-tipped flakers would still have had a positive effect on the flaking process and made the manufacture easier. The force needed to remove flakes is less, when copper is used. Likewise, flakes can be spaced more evenly, and the removals tend to be more regular, meaning that the lateral edges of the flakes tend to be more straight and even, which would be a highly desired feature for the aesthetic parallel retouch [66]. Though we have as yet no archaeological support for the existence of copper tips for flint manufacture in Scandinavia, it is not too far-fetched to assume they were available. As argued above, copper and some metallurgical knowledge was present in Scandinavia at the time. A tip for a pressure flaker would not require much material, and not much metallurgical knowledge would be needed to fabricate a short pin which can be inserted into a wooden handle. Likewise, in other European regions, copper objects exist, which suggests a connection and use in flint knapping contexts [68–70]. The possibility is therefore high that copper flakers were either not recovered archaeologically because the short pieces of copper were not preserved or were not recognised either due to heavy oxidation or because they were mistaken for other artefacts, like awls. Modern knappers use their copper-tipped pressure flakers for several years before the material is worn down so irrecoverably that it has to be replaced by a new tip. The demand for copper in the flint knapping context in prehistoric Scandinavia was thus quite likely very low, and resupplying would not have posed a major problem. The identification of copper in flint knapping processes in Scandinavia thus relies on technical observations and trace element analysis, which both have come up with heavily debated results, e.g. [68, 71–73].

3. Technical identification of copper in flint knapping

The data used in this chapter are part of a recent project concerning Late Neolithic and Early Bronze Age bifacial production [67]. For the study, production flakes from modern knappers replication of bifacial sickles and daggers were recorded and analysed, as well as their (i.e., the flint knappers) decision process (**Figure 4**). Questions about individual technical choices and transmission of knowledge were addressed. A minor part of the project was also concerned with the identification of copper in the knapping process. For this, not only technical markers were analysed, but some flint flakes were examined by scanning electron microscopy (SEM). A problem when trying to confirm traces on object surfaces by SEM is the limited size an object can have to be placed in the chamber [72, 74]. Likewise, it is time-consuming when no exact spot is known, and the entire surface has to be analysed. Part of the project was, therefore, to have a closer look at production flakes, not only to identify where copper implements leave traces of metal but also to ascertain which other attributes and characteristics are connected to the use of copper and thus can give hints in the absence of visible proof. Visible denotes here traces that are identifiable with the naked eye or a magnifying glass.

Flint is a generic term for silicious material with certain properties. It is mostly fine-grained, homogeneous and knappable [45, 75, 76]. Depending on the variety and the object that is being made, no further treatment of the flint is necessary. If tough and coarse-grained flint is used, heat treating may help to improve the knapping quality so flakes can be removed more easily [77, 78]. Heat treatment of flint has not been confirmed in Neolithic or Bronze Age contexts in Scandinavia. This is probably due to the generally good quality of available flint as well as the abundance

Figure 4.
Finished bifacial objects of the inventories included in the recent project that is referred to here. Outer left: Type IC dagger. Rest: Early bronze age type asymmetric sickles. © a. Heitman, UFG Kiel.

of the raw material [79]. When force is applied punctually to the material, it fractures conchoidal. The more homogeneous and isotropic the material, the more predictable the process is [80, 81]. The force applied results in a fracture following the Hertzian cone. The resulting crack in the material will bend inward rather quickly, and the force will run more or less parallel to the surface until the flake detaches completely. A second mechanical property used in flint knapping is the bending fracture. Without a clear point of impact, the stress on the material tears the flake away from the nodule [81, 82]. Both types of fracture mechanics leave characteristic traces on the flakes, which help to determine which fracture happened during the reduction. For example, a bulb is a sign of conchoidal fracture, while a lip formation indicates bending of the material. The bulb is the partial expression of the Hertzian cone, where the force of the blow has travelled through the material [82], while the lip is an overhang, left on the back edge of the flake platform. Further, the markedness and joint occurrence of certain technical attributes make it possible to identify which techniques were applied during the reduction process.

The first distinction in knapping techniques is drawn between direct and indirect application. This differentiates if the knapping tool was applied directly to the nodule or if an intermediate piece was placed between nodule and percussor. Classically, direct percussion is separated in hard percussion with a stone and soft percussion with organic implements, like antler or wood [45, 55, 83]. Depending on the knapping material, the flake detaches either by conchoidal or bending fracture. Generally speaking, the softer the material, the bigger the area of impact, and bending fractures are more likely to happen.

Another technique applied in a direct manner is pressure. During pressure flaking, a device, often a short handle with a pointed tip, is placed directly on the edge, and pressure is applied until the flake detaches. Pressure can also be applied with the so-called Ishi-stick. The principle is the same, but the Ishi stick is most often a rod of forearm length. Due to this, more leverage can be applied as the whole upper body can be used to exert pressure. The length of pressure flakers is not standardised and depends on the knapper and what they feel comfortable with. The detachment of the flake can happen by both conchoidal or bending fractures. Like direct percussion, pressure flaking can be split into a hard and a soft variety. Former is done, by using copper-tipped pressure flakers, while the latter uses tips of antler, wood, or bone [45, 84]. Indirect technique includes the use of an intermediate piece positioned on the edge where the flake is wished to be removed, on which the blow is then dealt. Like the pressure technique, the control is higher than in free percussion, as the exact spot for removal can be chosen, but the force delivered is higher than achieved by pressure. Intermediate pieces are called punch and are usually made from antler or copper, but other materials are possible [45].

As mentioned above, the choice of technique can be reconstructed by analysing technical markers on the flakes. Relevant to this chapter are only characteristics related to knapping with copper tools, so a further discussion of attributes related to other implements will not be given. It should be noted, though, that the overlap in attributes is very high, and they are not mutually exclusive [81]. Statements about the production process should, therefore, be based on a multitude of flakes being examined and not on single observations.

Characteristics of pressure flaking are small and plain platform remnants without facets, barely extending the contact area of the implement tip. The harder the tip's material, the smaller the remnant can be, and the more likely conchoidal fracturing becomes [81, 84, 85]. The bulb is often short and set rather high on the ventral surface (the surface that was attached to the nodule), and if copper is used for pressure, very small ring cracks are possible [81, 85, 86]. Ring cracks are small, semi-circular fissures indicating the top of the Hertzian cone [87]. During the recording of production flakes, it was noted that flint flakes with visible confirmation of copper traces indeed had quite marked but small and high-set bulbs. However, this was not true in every case, which can be explained by the state of preservation of the knapping device. Regardless of the technique used, the continuous impact of the implement on the more or less sharp edge of flint wears down the material. An important part of flint knapping is thus the maintenance of the knapping implement's surface. During pressure flaking, no blow is dealt, but a lot of force is exerted on a rather small area, which also wears down the tips of antler as well as of copper. For the maximum effect, the tips have to be kept in good working order and sharpened every once in a while. For copper, this can happen through grinding or hammering. The latter has the advantage of hardening the material further, as it resembles cold forging. If the tip wears down and the area of contact with the material gets slightly bigger, this has effects on the knapping attributes; the less sharp the point is, the less marked the attributes appear [88], which explains why traces are visible, but the attributes do not match in every case.

Most research concerning attribute formation in copper flaking has been done in connection to blade production, e.g. [86, 89, 90]. Occasionally, copper flaking has been identified in contexts of pressure flaking, e.g. [72, 91], but the focus of analysis was always on finished objects and not production flakes. Experimental production of artefacts has shown that copper-tipped flakers leave visible traces when the tip slips and scratches across the surface, which often also happens when the termination fails, and the tip connects to the resulting step in the material [72]. For the experimental replications in this study, all craftsmen were allowed to use their own tools at their own discretion. As copper pressure flakers cannot be confirmed or excluded from the production process archaeologically, this had no influence on the authenticity of the replications, even if there is no evidence for the use in sickle production during prehistory. All resulting production sequences were documented and compared to each other, and an ideal sequence for bifacial dagger production [67]. Compared to the ideal sequence, copper was implemented and used more often and earlier in the process than expected. The choice to use copper pressure flakers was also dependent on the individual knapper's preferences. While two of the knappers preferred to work with copper to have more control over of the process and be able to progress more carefully, the third preferred not to use copper because it tends to leave deeper negatives, which have to be compensated later.

Visual inspection of the finished artefacts (**Figure 4**) did yield traces of copper on the surface, mostly in connection with failed terminations, but not on all pieces and surprisingly not on the included type IC dagger replication, where the last production step is executed with copper pressure flaking alone. The confirmation of copper traces left on production flakes, especially on the platform remnants, was better. Out of 12,184 recorded flakes, 942 had visible traces of copper, mostly on the platform remnant or below the platform on the edge (**Table 1**). The dagger replication by AB especially included a very high number of traces, which is probably due to the relatively poor quality of the raw material, which forced the knapper to progress carefully. In his case, this included the default to copper pressure but also the implementation of copper punches during indirect percussion, which was chosen more often for safety reasons.

As mentioned above, ring cracks are used as a representation for flint knapping with copper in the absence of other evidence [86, 91]. The recorded ring cracks from the case study were also analysed to see if any prediction concerning the use of copper implements in the knapping process could be made. **Table 2** shows the expected low frequency of ring cracks, as most of the production is executed in soft percussion. Most of the cracks are probably related to pressure and indirect technique, but they do not seem to be the only factor. As mentioned already, AB used copper punches rather frequently during the production of the dagger. This was not done in the sickle production, and pressure with copper was not done extensively. Still, the frequency

Artefact	Count with copper	Total number recorded
AB dagger	670	3330
AB sickle	9	1730
GN dagger	9	2109
GN sickle 2006	89	1059
GN sickle 2007	158	1771
PW sickle 2006	1	997
PW sickle 2007	6	1081
Total	**942**	**12,077**

Differing total number of flakes from the total amount mentioned in the text is due to an additional test inventory, which has not been included in the paper.

Table 1.
Number of flakes with visible copper traces on the surface in each recorded inventory.

Inventory	None	Present	With conical break	Not preserved
AB dagger	84.6	4.84	0.00	10.6
AB sickle	56.1	8.66	0.18	35.0
GN dagger	69.2	13.86	0.14	16.9
GN sickle 2006	75.3	8.03	0.09	16.5
GN sickle 2007	83.7	3.06	0.06	13.2
PW sickle 2006	65.8	8.68	0.31	25.2
PW sickle 2007	70.1	5.98	0.00	23.9

Table 2.
Frequency in percent of ring cracks in the recorded inventories.

of ring cracks is nearly double in the sickle inventory. AB stated that he uses copper when he wants to be more in control of the knapping process and use less force during the removal. Thus, choosing the harder material here seems to translate to less forceful flaking, which reduces the formation of ring cracks. In contrast, GN dagger shows a rather high percentage, which relates back to the raw material. This inventory is the only one using differing materials: Texas flint, which is a tougher and less homogeneous variety compared to Scandinavian flint. More force is thus needed to remove flakes, which favours the formation of ring cracks.

Further analysis showed that ring cracks appear mostly in the first stages of production and not so much in the last. Only a few have been recorded on the smallest flakes, which are more likely to originate from the final pressure retouch or on the flakes from the parallel retouch (**Figure 5**). This strengthens the impression that

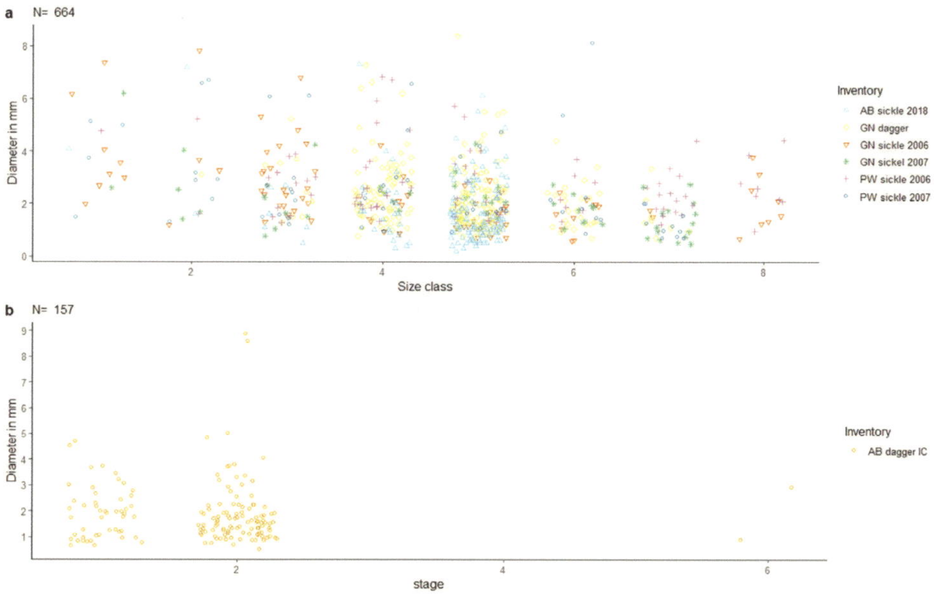

Figure 5.
Scatterplot of ring crack diameters by production stage or flake size class. Ring cracks exceeding 10 mm were removed from the data sets. The flake size classes used here refer to the length of the flake. Size class 1 comprises the biggest flakes (≥ 100 mm) and the length gradually declines until flake class 8 (5–6.99 mm).

it is the indirect technique which is responsible for the ring cracks in the inventories. No, or just a few ring cracks during the parallel flaking and pressure retouch is not so surprising, as the amount of force needed is rather small compared to the earlier stages, where more and larger areas of the surface are being removed. This shows that the force needed to remove the parallel flakes is not high enough to leave ring cracks in a repetitive way, and the attribute is not a significant indicator for the use of copper-tipped pressure flakers in the archaeological record if the bifacial method is analysed. This draws attention to a problem relating to the experimental identification of knapping attributes. As mentioned, the attributes for copper flaking have been identified in experiments of blade production [85, 86, 89, 92]. Blades, and especially blades knapped with copper, are long, narrow, and regular flakes. They need a certain kind of preparation, precision, and force to be removed from the nodule. The conditions for pressure flaking or indirect percussion in a bifacial method differ completely. It is shown here that the attributes are not transferable one-to-one to other production strategies.

To get a better impression how the ring crack diameter relates to pressure flaking and indirect technique using copper implements, all flakes with visible traces were assembled, and the ring crack diameters were plotted against the distance from the edge (**Figure 6**). This was done to see if distinct groups would be observable. Indirect technique is mostly used in the early stages of production and is meant to remove larger flakes, and as the punch is bigger than the pressure flaker tip, this means the flake is removed at a greater distance from the edge than a pressure flake. As **Figure 4** shows, no group formation can be detected. As AB was the only knapper including copper punches in the production process and only during the production of the dagger, all other inventories and plotted points indicate that pressure flaking with copper indeed relates to small ring cracks near the edge. But, the relatively low number of incidences indicates that other factors need to be met so that ring cracks really appear.

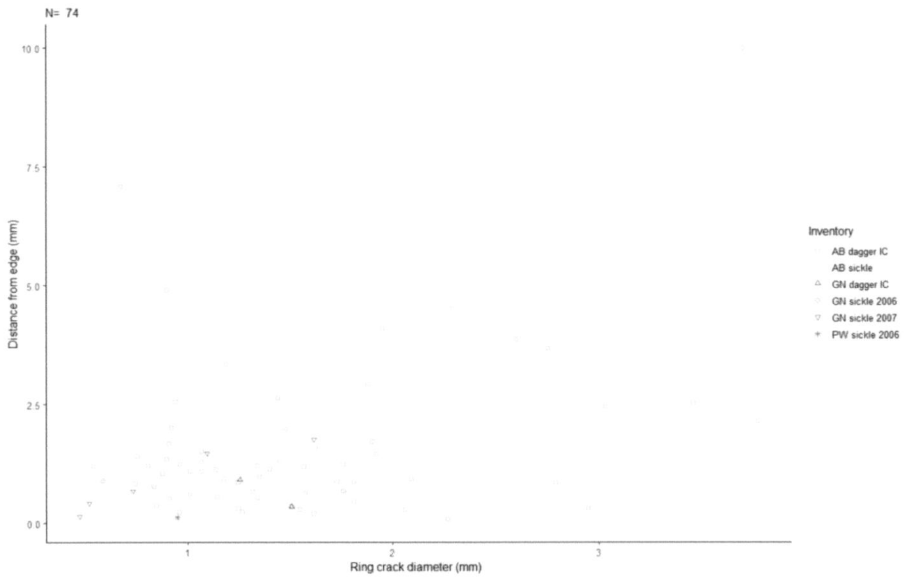

Figure 6.
Scatterplot of all ring cracks on flakes with visible confirmation of copper traces in correlation to the distance of the impact point from the edge of the platform remnant.

Likewise, no clear separation between pressure and indirect technique with copper seems to be present, at least not in the formation of ring cracks.

Thus, even if it is known when and why copper was included in the production process, this does not mean an identification is easy. **Table 1** shows that copper has an ambiguous tendency to leave traces on the surface of flint. The frequency of implementation, skill of the knapper, force needed for the removal, as well as raw material qualities are all aspects, which have an influence on the possibility of copper traces being visible on the surface. Likewise, the copper itself is important; the softer the material, the more likely it will leave traces. Hardened tips are thus more likely not to leave visible traces. Also, the handling by the knapper has influence; skilled and experienced knappers are less prone to errors and accidents during the reduction. This lessens the chance of scratches on the surface by slipping. Further, the technical markers and attributes are not always explicit.

The possibility of detecting traces on the platform remnant of flakes would thus offer a higher chance of identifying knappers working with copper, as even the experienced knappers need to place the tip on the surface of the flint and exert pressure for the detachment. As mentioned above, interpretation of used techniques should not be made on single flakes, as the characteristics overlap, and clear-cut evidence for copper in the knapping process only shows on very few flakes. It would be useful to have a more reliable way to identify copper-knapped flakes, and trace element analysis can be a method for that. Using flakes for the detection of copper in flint knapping processes offers possibilities the finished tools cannot offer. Flakes are mostly not subjected to further use, which, especially for the sickles, can remove traces left on the surface. Likewise, the possible areas to look for traces are smaller and thus more quickly inspected, even in the absence of visual evidence.

4. Material analysis and results

Residue analysis on flint started to develop in the 1960s, following S. Semenov's [93] seminal work on flint technology [74]. Studies like P.C. Anderson's [94] work showed that residues of different kinds leave traces on the flint and can be used to identify the material which was worked. Most often, residue analysis is done to make statements about the function and handling of flint tools, e.g. [95–97]. Studies relating to the identification of knapping tools are scant, which relates back to the facts that use-wear traces need time and repetition to form – which is not given during knapping, as the knapping tool only connects briefly with the surface – and organic residue on very small areas, as the platform remnant, quite likely have perished by the time archaeologists uncover the artefacts. Likewise, even modern residues can be hard to interpret visually, as their morphology can vary depending on many factors other than degradation [74].

Though residue and use-wear analysis have become an integral part of archaeology, no real best practice guide or norm of analysis has been formulated yet [74].

Some flakes with visible confirmation of copper traces from experimental replicas were selected for trace analysis. They were each subjected to visual analysis under the Andonstar ADSM302 digital microscope, and later, the chemical analysis was conducted using SEM Zeiss Gemini Ultra55 Plus. The SEM uses radiation (secondary electrons and X-rays) created by the interaction of the primary electrons and the sample in order to receive information about surface morphology and chemical composition of the same micro-scale area. For the imaging, secondary electrons with relatively low energy provide the most surface-sensitive contrast, so-called SE-imaging. Compared to light microscopy, SE-imaging provides a much higher depth of field, giving surfaces the impression of three-dimensional micro-topographies. However, due to the conventionally used electron detectors, images are restricted to greyscale. Localisation of residue can thus be quite challenging, which is why initial analysis with visual microscopes is recommended [74]. The instrument applied here uses a beam of electrons operating in the accelerating voltage of 5–20 kV. The electron images recorded via SEM were used to attain the surface topography of the specimen, showing the contrast between the possible metal traces on the flint material. Furthermore, chemical analyses via elemental mapping were conducted in the energy-dispersive X-ray (EDX) mode in SEM. In this case, the beam of electrons scans the surface of the specimen, and secondary X-rays are detected by the EDX detector to determine the chemical composition laterally resolved. In the scope of this study, the EDX point measurements and elemental mapping have been conducted. The X-ray spectrum always includes weak stray signals from the instrument itself, the sample holder, etc., which has to be kept in mind when interpreting the complete spectrum [74]. Furthermore, non-conducting artefacts had to be coated with carbon

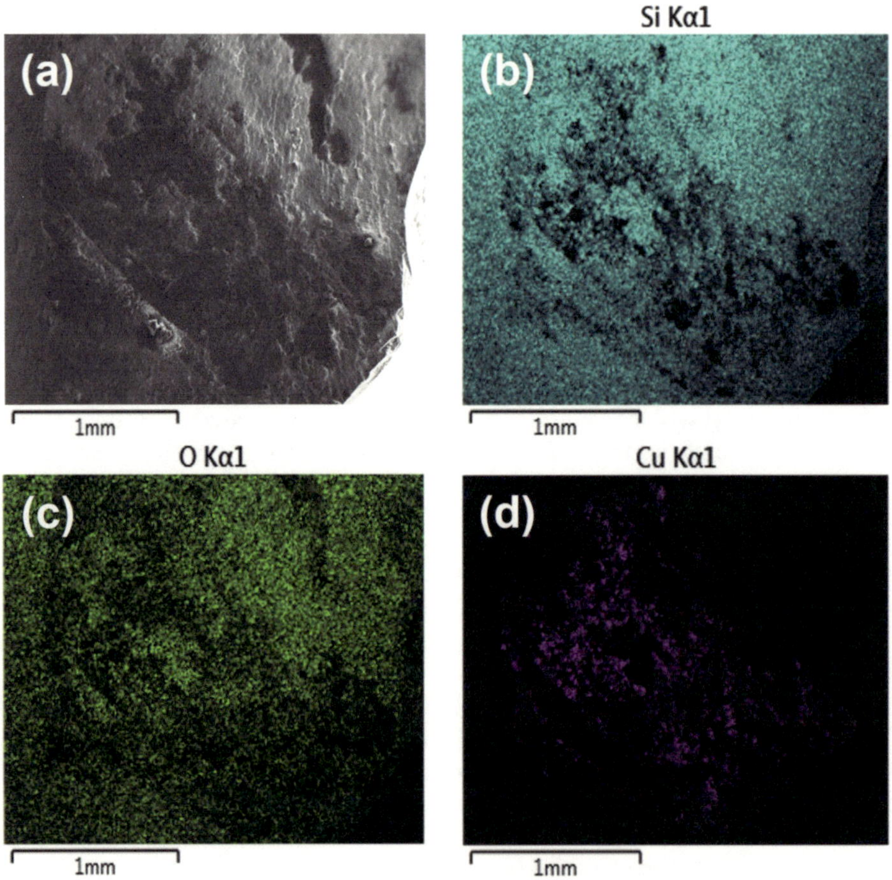

Figure 7.
SEM EDX elemental mapping on the flake 761 with electron image (a) and images (b-d) showing the corresponding elements along with the the copper traces.

or gold for high-quality imaging. However, more recent studies have shown that it is possible to collect images and spectra without coating the surfaces [74, 98].

Two representative examples of the analyses are shown here. In **Figure 7**, a flake (number 761) produced by AB from the replication of an Early Bronze Age flint sickle is displayed and records the observations from the SEM. In the electron image, certain regions appear brighter, corresponding to the original flaked surface of the flint, which contains Si and O as determined by the elemental mapping. Moreover, the EDX elemental mapping also shows the signals from the copper metal, which are residues from the copper knapping tool. Similarly, **Figure 8** also shows the traces of copper as confirmed by the SEM EDX elemental mapping showing the copper metal traces on the surface of the replication of the flint 1609.

In another experiment, the flakes were heated at 600 °C for 24 and 48 hours to simulate oxidation. The result was that the copper traces on the surface do reduce through time (**Figure 9** and **Table 3**) by the thermal treatment; however, the mechanical stability of the copper oxide formed is sufficient to avoid complete removal from the surface. The same is presumably true for flint, which has been exposed to the environment for a long time. Consequently, the confirmation of copper traces by SEM-EDX does not provide an unambiguous indication for the use of copper tools in the production process. Likewise,

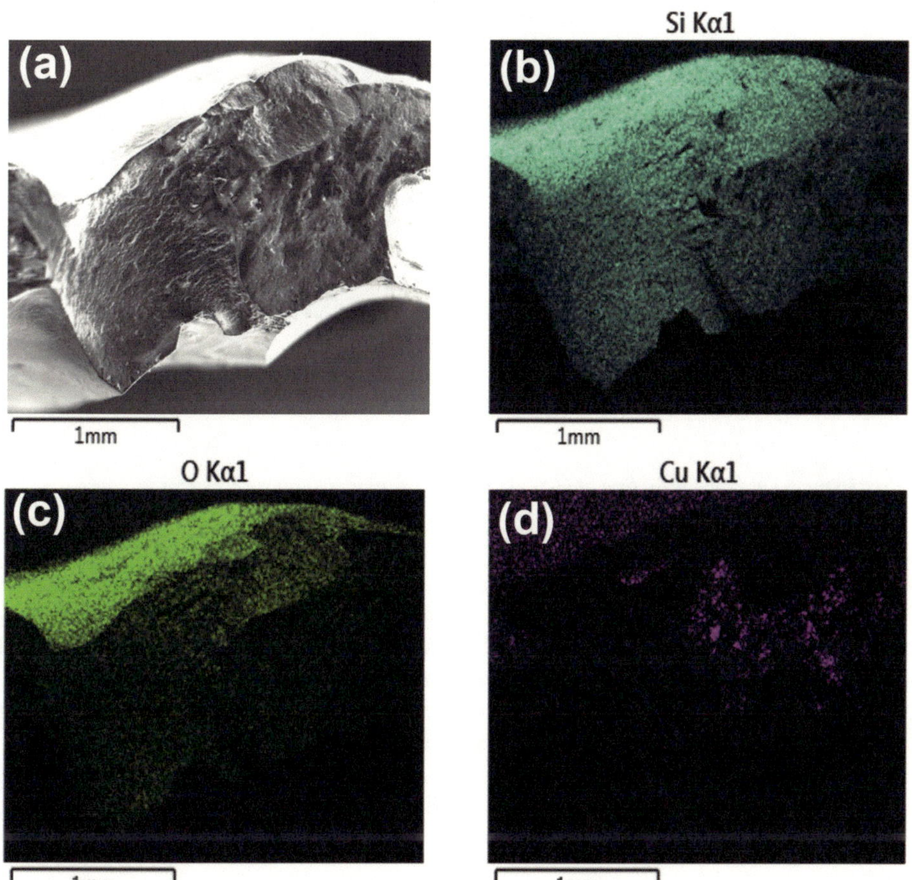

Figure 8.
SEM EDX elemental mapping on the flake 1609 with electron image (a) and images (b-d) showing the corresponding elements along with the the copper traces.

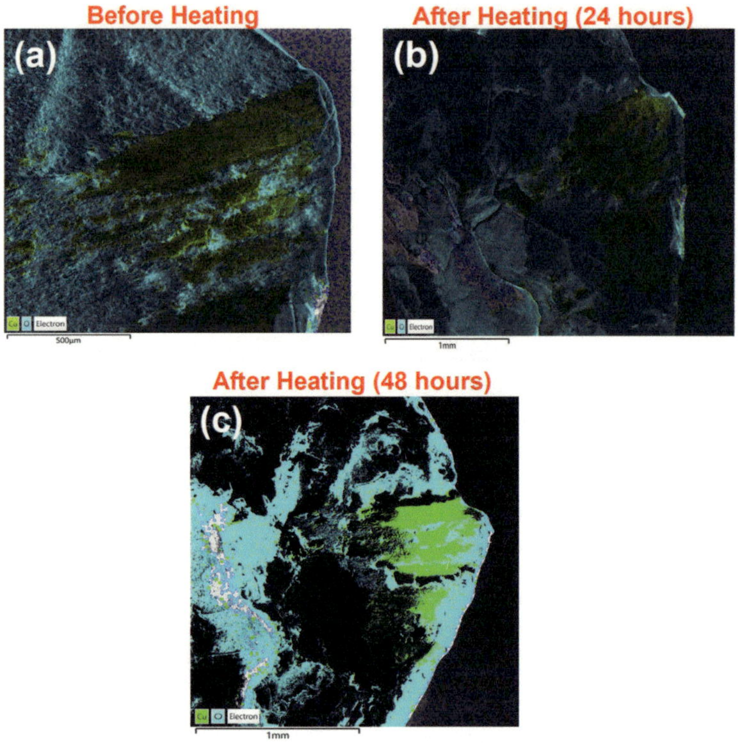

Figure 9.
SEM EDX elemental map of experimental flint replica (a) before heating, (b) after heating for 24 hours, and (c) after heating for 48 hours.

Atomic percent (at.%)	Before heating	After heating 24 hours	After heating 48 hours
Cu	76 at.%	36 at.%	17 at.%
O	23 at.%	64 at.%	83 at.%

Table 3.
SEM electron images of heat treated flint at 600 °C (a) before heating, (b) after heating for 24 hours, and (c) after heating for 48 hours.

only examining technical attributes is not sufficient and may lead to wrong conclusions. Further analytical strategies should be based on well-designed combinations of correlated analytical techniques. For instance, dedicated (e.g. Synchrotron-based) XRF and/or Synchrotron-based tomography may provide information about the presence of copper. Using identical location approaches, the chemical nature of the copper (oxide, element, etc.) may be determined by SEM, particularly if the measurements are already supported by machine learning approaches. The ultimate test for such a novel strategy would be to identify the nature of copper traces on flint without visible indication.

5. Discussion

Identification of copper pressure flakers or punches in the production process of flint objects is not as straightforward, as often described especially in technological studies. As could be shown in the flake inventories included here, the technical attributes identified as typical markers for the involvement of copper during blade production are not present or are not only related to copper in the flaking process. The very small ring crack, which has mostly been used to identify copper pressure flakers, is mostly absent on the production flakes analysed. Or the ring cracks associated with copper traces have a bigger diameter than assumed in connection. Likewise, markers such as the bulb, which can be quite distinctive, do not always show the same features when knapping with copper. The state of the copper tip as well as the hardness of the material or force of application can obscure the distinct features, so a definite assignment is not possible.

Likewise, trace element analysis has only been done in connection with finished artefacts so far, which poses several problems. First, the size of the objects for which analyses are possible. Second, there is a concentration on few and mostly finely made objects, in this case also only daggers from Bronze Age contexts, e.g. [71, 72]. Confirming copper in contexts from which we know copper had to be present to achieve the outcome is a bit of a self-confirming result. It tells us nothing new but still fails to answer if copper as knapping tool was widely available or only restricted to very few craftspersons. Likewise, as could also be seen in this study, when skilled craftspersons are involved in the process, there does not need to be traces left on the surface. They are mostly connected to errors in the knapping process and can also be removed from the surface again when those errors are eliminated, e.g. [72]. There is also the possibility for accumulation of element traces due to depositional circumstances, e.g. [99], as well as removal through oxidation. Mostly, if copper and other materials had sufficient contact with the flint surface to leave elemental traces, those remains would seem to be amply preserved to be detectable in analysis, even after long years in the soil [72, 74], but it has to be located before analysis nonetheless. Also, no studies exist so far that deal with the question of how use affects the traces. The study by G.H. Strand Tanner [72] included only a very limited experiment on cutting flesh.

Looking for traces of production flakes is therefore the next logical step. They are mostly not restricted by the size requirements, especially as flakes removed by copper pressure tend to be relatively small. Likewise, the area of interest is so much smaller, i.e. mostly relating to the platform remnant and the edge below, though steps and hinges of failed terminations can be of interest, too. Another advantage is the certainty that if copper was involved, the tip had to connect to the platform. On the surface of finished artefacts, every trace detectable is an impact by chance, contrary to the impact point on the platform remnant. If the knapping device would not have connected, the flake would not have detached. This simplifies and shortens the analysis with the SEM a lot. Furthermore, in combination with technical markers, flakes can be pre-sorted for probability of bearing traces, which can reduce the negative results for a time-consuming and expensive analysis.

6. Conclusion

Copper in the knapping process can best be identified by combining approaches. Trace element analysis can confirm the presence of copper traces on the surface but works best when distinct locations for sampling are known. These can be found either by visual confirmation or by selecting technical markers. A problem with the bifacial method is that the markers identified so far are not represented very well in the production process. This is most likely due to the differences in production, as the previously published experiments were conducted with blade manufacture, which is very dissimilar to the bifacial method. Further experiments of copper implementation in bifacial production would be needed to distinguish typical characteristics on flakes. If those are identified, it will help to answer a wider variety of questions concerning not only the technological knowledge of flint knapping but also metallurgy. Also, questions about social developments could be pursued, for example, in the context of the availability of copper to flint knappers, and this also gives insight into social networks of dependency.

Likewise, identifying the implementation of copper tools in flint production processes can help us to better describe the development of flint knapping technology and understand how the elaboration of the craft was structured. There is no doubt that metal and metallurgical knowledge influenced flint technology, not only in the form and function of objects but also in the manufacturing process. Copper offers advantages compared to antler or wood and eases the process of knapping while allowing for more control. This is true not only for production steps, which are not possible without copper. Many modern flint knappers rely a lot on copper during flint knapping, depending on how authentic they want to work. There are technically no reasons why prehistoric knappers should not have done the same, when copper became available, save social and traditional rules structuring the manufacturing process. If such rules were in existence and/or how they changed, the prospect of changing technologies can be explored better if we understand how the new techniques were integrated into existing technologies. The effect is visible in the typological and technical development of the flint daggers, but when, where, and how the development started is still not fully understood. Analysing production flakes in favour of finished tools offers a more holistic and objective approach to the questions.

Acknowledgements

This work was funded by the Deutsche Forschungsgemeinschaft (DFG, German Research Foundation) under Germany's Excellence Strategy-EXC 2150-390870439.

Conflict of interest

The authors declare no conflict of interest.

Author details

Moiken Hinrichs[1*], Khurram Saleem[2], Berit V. Eriksen[1] and Lorenz Kienle[2]

1 Cluster of Excellence ROOTS, Kiel University, Kiel, Germany

2 Department of Materials Science, Synthesis and Real Structure, Cluster of Excellence ROOTS, Kiel University, Kiel, Germany

*Address all correspondence to: moikenhinrichs@web.de

References

[1] Semaw S. The world's oldest stone artefacts from Gona, Ethiopia: Their implications for understanding stone technology and patterns of human evolution between 2.6-1.5 million years ago. Journal of Archaeological Science. 2000;**27**(12):1197-1214

[2] Harmand S, Lewis JE, Feibel CS, Lepre CJ, Prat S, Lenoble A, et al. 3.3-million-year-old stone tools from Lomekwi 3, West Turkana, Kenya. Nature. 2015;**521**(7552):310-315

[3] Bello SM, Delbarre G, De Groote I, Parfitt SA. A newly discovered antler flint-knapping hammer and the question of their rarity in the palaeolithic archaeological record: Reality or bias? Quaternary International. 2016;**403**:107-117

[4] Clément S. Soft-hammer percussion during the Acheulean: Barking up the wrong tree of technical change? Journal of Paleolithic Archaeology. 2022;**5**(1):3. DOI: 10.1007/s41982-021-00104-6

[5] Reingruber A. Absolute and relative chronologies in the lower Danube area during the 5th millennium BC. In: Hansen S, Raczky P, Anders A, Reingruber A, editors. Neolithic and Copper Age between the Carpathians and the Aegean Sea Chronologies and Technologies from the 6th to the 4th Millennium BCE International Workshop Budapest 2012. Bonn: Dr. Rudolf Habelt GmbH; 2015. pp. 301-324

[6] Brozio JP, Stos-Gale Z, Müller J, Müller-Scheeßel N, Schultrich S, Fritsch B, et al. The origin of neolithic copper on the central northern European plain and in southern Scandinavia: Connectivities on a European scale. PLoS One. 2023;**18**(5):e0283007

[7] Roberts BW, Frieman CJ. Early metallurgy in Western and northern Europe. In: Fowler C, Harding J, Hofmann D, editors. The Oxford Handbook of Neolithic Europe. Oxford: Oxford University Press; 2009. pp. 711-728. DOI: 10.1093/oxfordhb/9780199545841.013.037

[8] Scharl S, Denis S, Collin JP, Allard P, Delvigne V, Gehlen B, et al. Studying neolithic lithics – From a cross-border dialogue to a common language. Journal of Neolithic Archaeology. 2021;**23**:115-128

[9] Frieman CJ. An Archaeology of Innovation: Approaching Social and Technological Change in Human Society. Manchester: Manchester University Press; 2023. 238 p

[10] Roberts BW. Migration, craft expertise and metallurgy. Analysing the 'spread' of metal in Western Europe. Archaeological Review From Cambridge. 2008;**23**(2):27-45

[11] Ottaway BS. Innovation, production and specialization in early prehistoric copper metallurgy. European Journal of Archaeology. 2001;**4**(1):87-112

[12] Amicone S, Radivojević M, Quinn P, Rehren T. Pottery technology at the dawn of metallurgy in the Vinča culture. In: Radivojević M, Roberts BW, Marić M, Kuzmanović Cvetković J, Rehren T, editors. The Rise of Metallurgy in Eurasia Evolution, Organisation and Consumption of Early Metal in the Balkans. Oxford: Archaeopress Publishing Ltd; 2021. pp. 538-551

[13] Kienlin TL. Tradition and innovation in copper age metallurgy: Results of a metallographic examination of flat axes

from eastern Central Europe and the Carpathian Basin. Proceedings of the Prehistoric Society. 2008;**74**:79-108

[14] Scharl S. Patterns of innovation transfer and the spread of copper metallurgy to Central Europe. European Journal of Archaeology. 2016;**19**(2):215-244

[15] Earle T, Ling J, Uhnér C, Stos-Gale Z, Melheim L. The political economy and metal trade in bronze age Europe: Understanding regional variability in terms of comparative advantages and articulations. European Journal of Archaeology. 2015;**18**(4):633-657

[16] Vandkilde H. From stone to bronze: The metalwork of the late neolithic and earliest Bronze Age in Denmark [thesis]. Aarhus: Jutland Archaeological Society Moesgaard; 1996

[17] Manclossi F, Rosen SA, Boëda E. From stone to metal: The dynamics of technological change in the decline of chipped stone tool production. A case study from the southern Levant (5th–1st millennia BCE). Journal of Archaeological Method and Theory. 2019;**26**:1276-1326

[18] Eriksen BV, editor. Lithic Technology in Metal Using Societies: Proceedings of a UISPP Workshop, Lisbon, September 2006. Højbjerg: Jutland Archaeological Society; 2010

[19] Kristiansen K, Larsson TB. The Rise of Bronze Age Society: Travels, Transmissions and Transformations. Cambridge: Cambridge University Press; 2005. 464 p

[20] Klassen L. Jade und Kupfer: Untersuchungen zum Neolithisierungsprozess im westlichen Ostseeraum unter besonderer Berücksichtigung der Kulturentwicklung Europas 5500-3500 BC. Højbjerg: Jutland Archaeological Society; 2004. 435 p

[21] Becker CJ. Late-neolithic flint mines at Aalborg. Acta Archaeologica. 1951;**22**:135-152

[22] Becker CJ. Flint mining in neolithic Denmark. Antiquity. 1959;**33**(130):87-92

[23] Ling J, Hjärthner-Holdar E, Grandin L, Billström K, Persson PO. Moving metals or indigenous mining? Provenancing Scandinavian bronze age artefacts by lead isotopes and trace elements. Journal of Archaeological Science. 2013;**40**(1):291-304

[24] O'Brien W. Prehistoric Copper Mining in Europe, 5500-500 BC. 1st ed. Oxford, New York: Oxford University Press; 2015. 345 p

[25] Olausson D. The Flintknapper and the Bronzesmith. In: Bergerbrant S, Sabatini S, editors. Counterpoint: Essays in Archaeology and Heritage Studies in Honour of Professor Kristian Kristiansen. Oxford: Archaeopress; 2013. pp. 447-456

[26] Nørgaard HW, Pernicka E, Vandkilde H. Shifting networks and mixing metals: Changing metal trade routes to Scandinavia correlate with neolithic and bronze age transformations. PLoS One. 2021;**16**(6):e0252376

[27] Gebauer AB, Sørensen LV, Taube M, Wielandt DKP. First metallurgy in northern Europe: An early neolithic crucible and a possible Tuyère from Lønt, Denmark. European Journal of Archaeology. 2021;**24**(1):27-47

[28] Magnusson SB. Early metallurgy in Scandinavia: Continuity or discontinuity? Lund Archaeological Review. 2001;**7**:19-28

[29] Kuijpers MHG. An Archaeology of Skill: Metalworking Skill and Material Specialization in Early Bronze Age Central Europe. 1st ed. London: Taylor and Francis; 2018

[30] Skorna H. The Life and Journey of Neolithic Copper Objects. Transformations of the Neuenkirchen Hoard, North-East Germany (3800 BCE). Leiden: Sidestone Press; 2022. 198 p

[31] Missed MJ, Innovation. The earliest copper Daggers in northern Central Europe and southern Scandinavia. In: Bergerbrant S, Sabatini S, editors. Counterpoint Essays in Archaeology and Heritage Studies in Honour of Professor Kristian Kristiansen. Oxford: Archaeopress; 2013. pp. 443-448

[32] Radivojević M, Roberts BW, Pernicka E, Stos-Gale Z, Martinón-Torres M, Rehren T, et al. The provenance, use, and circulation of metals in the European bronze age: The state of debate. Journal of Archaeological Research. 2019;**27**(2):131-185

[33] Kristiansen K. The rise of bronze age pereipheries and the expansion of international trade 1950-1100 BC. In: Kristiansen K, Lindkvist T, Myrdal J, editors. Trade and Civilisation: Economic Networks and Cultural Ties, from Prehistory to the Early Modern Era. Cambridge; New York: Cambridge University Press; 2018. pp. 87-112

[34] Serbe B. Bernstein in der Bronzezeit. In: Netzwerke und Interaktion in Europa [thesis]. Kiel: Christian-Albrechts-Universität zu Kiel; 2023

[35] Roberts BW. Production networks and consumer choice in the earliest metal of Western Europe. Journal of World Prehistory. 2009;**22**(4):461-481

[36] Roberts BW, Frieman CJ. Drawing boundaries and building models: Investigating the concept of the 'chalcolithic frontier' in Northwest Europe. In: Allen MJ, Gardiner J, Sheridan A, editors. Is there a British Chalcolithic? People, Place and Polity in the Later Third Millennium. Oxford: Oxbow Books; 2012. pp. 27-39

[37] Müller J. Funnel beaker societies and long-distance trade. In: Ling J, Chacon RJ, Kristiansen K, editors. Trade before Civilization Long-Distance Exchange and the Rise of Social Complexity. Cambridge; New York: Cambridge University Press; 2022. pp. 23-39

[38] Rosen SA. Lithics after the Stone Age: A Handbook of Stone Tools from the Levant. Walnut Creek, CA: AltaMira Press; 1997. 184 p

[39] Müller S. Flintdolkene i den nordiske stenalder. Nordiske Fortidsminder. 1902;**1**(4):125-180

[40] Forssander JE. Der ostskandinavische Norden während der ältesten Metallzeit Europas. Lund: Gleerup; 1936. 296 p

[41] Glob PV. Stenredskaber fra Bronzealderen. In: Norling-Christensen H, Glob PV, editors. Winther-Festskrift Til Købmanden Jens Winther Paa 75-Aarsdagen. København: Ejnar Munksgaard; 1938. pp. 40-74

[42] Frieman CJ. Flint daggers, copper daggers, and technological innovation on late neolithic Scandinavia. European Journal of Archaeology. 2012;**15**(3):440-464

[43] Vandkilde H. A review of the early late neolithic period in Denmark: Practice, identity and connectivity. Offa. 2007;**61/62**(2004/05):75-109

[44] Olausson D. Making daggers and scouting talents: Situated leaning in late neolithic Scandinavia. In: Johannsen N,

Jessen MD, Jensen HJ, editors. Excavating the Mind. Aarhus: Aarhus University Press; 2012. pp. 211-232

[45] Inizan ML, Reduron-Ballinger M, Roche H, Tixier J, Féblot-Augustins J. Technology and Terminology of Knapped Stone: Followed by a Multilingual Vocabulary - Arabic, English, French, German, Greek, Italian, Portuguese, Spanish. Nanterre: CREP; 1999. 189 p

[46] Olausson D. Does practice make perfect? Craft expertise as a factor in aggrandizer strategies. Journal of Archaeological Method and Theory. 2008;**15**(1):28-50

[47] Müller S. Ordning af Danmarks oldsager I. Sten- og Bronzealderen. Paris, København, London, Leipzig: Librairie Renouard, C. A. Reitzel, Williams & Norgate, F. A. Brockhaus' Sortiment; 1888

[48] Kühn HJ. Das Spätneolithikum in Schleswig-Holstein. Neumünster: Wachholtz; 1979

[49] Lomborg E. Die Flintdolche Dänemarks: Studien über Chronologie und Kulturbeziehungen des südskandinavischen Spätneolithikums. Koebenhavn: Lynge (in Komm.); 1973

[50] Apel J. Daggers, knowledge & power: The social aspects of flint-dagger technology in Scandinavia 2350-1500 cal BC [thesis]. Uppsala: Department of Archaeology and Ancient History, University of Uppsala; 2001. 365 p

[51] Madsen T. Perioder og periodeovergange i neolitikum. Om forskellige fundtypers egnethed til kronologiske opdelinger. Hikuin. 1978;**4**:51-60

[52] Olausson D. Knapping skill and craft specialization in late neolithic flint Daggers. Lithic Technology. 2017;**42**(4):127-139

[53] Stafford M. In search of hindsgavl: Experiments in the production of neolithic danish flint daggers. Antiquity. 1998;**72**(276):338-349

[54] Stenak S, Unmasking the Maker. A personal lithic study [thesis]. Skive: Arco; 2022

[55] Pelegrin J. Les techniques de débitage laminaire au Tardiglaciaire: critères de diagnose et quelques réflexions. In: Valentin B, Bodu P, Christensen M, editors. L'Europe centrale et septentrionale au tardiglaciaire: confrontation des modèles régionaux de peuplement. Nemours: Association pour la promotion de la recherche archéologique en Ile-de-France; 2000. pp. 73-86

[56] Hansen PV, Madsen B. Flint axe manufacture in the neolithic: An experimental investigation of a flint axe manufacture site at Hastrup Vænget, East Zealand. Journal of Danish Archaeology. 1983;**2**(1):43-59

[57] Agthe M. Bemerkungen zu Feuersteindolchen im nordwestlichen Verbreitungsgebiet der Aunjetitzer Kultur. Arbeits- und Forschungsberichte zur Sächsischen Bodendenkmalpflege. 1989;**33**:15-113

[58] Cassau A. Ein Feuersteindolch mit Holzgriff und Lederscheide aus Wiepenkathen, Kreis Stade. Mannus Zeitschrift für Deutsche Vorgeschichte. 1935;**27**(1):198-209

[59] Fischer Mortensen M, Bennike O, Jensen LE, Jessen C, Juul A, Hjelm Petersen A, et al. Fortidens spor og fremtidens forbindelse – bevaring og naturvidenskab på Femern Bælt projektet. Danmarks største arkæologiske

udgravning. Nationalmuseets Arbejdsmark. 2015;**2015**:22-35

[60] Flintdolk fra ældre bronzealder [Internet]. Museum Lolland-Falster. 2014. Available from: https://web.archive.org/web/20141008165423/https://aabne-samlinger.dk/femernforbindelsen/femernforbindelsen/nyheder/2014/141006_flintdolk/ [Accessed: December 17, 2023]

[61] Museum Lolland-Falster. Enestående pragtvåben fundet ved Rødbyhavn. Flintdolk fra ældre bronzealder med barkskæfte er fundet for første gang i Danmark [Internet]. 2014. Available from: https://web.archive.org/web/20141020184049/http://www.aabne-samlinger.dk/femernforbindelsen/femernforbindelsen/arkaeologi/fund-fra-udgravningerne/flint/flintdolk-fra-aeldre-bronzealder/ [Accessed: July 2, 2023]

[62] Kersten K. Vorgeschichte des Kreises Steinburg. Neumünster: Wachholtz; 1939. 490 p

[63] Neolithic CE, Daggers D, Vol. II: A Manual for Flintknappers & Lithic Analysts. 1st ed. Lynchburg, VA: Warwick House Publishing; 2016

[64] Brøndsted J. Danmarks Oldtid. I Stenalderen. 2nd ed. København: Gyldendal; 1957. 375 p

[65] Rassmann K. Spätneolithikum und frühe Bronzezeit im Flachland zwischen Elbe und Oder. Lübstorf: Archäologisches Landesmuseum für Mecklenburg-Vorpommern; 1993. 359 p

[66] Stafford M. The parallel-flaked flint daggers of late neolithic Denmark: An experimental perspective. Journal of Archaeological Science. 2003;**30**(12):1537-1550

[67] Hinrichs M. Craftful Minds - Identifying Technical Individuality in Production Processes [thesis]. Kiel: Christian-Albrechts-Universität zu Kiel; 2023

[68] Nicolas C. Arrows of power from Brittany to Denmark (2500-1700 BC). Proceedings of the Prehistoric Society. 2017;**83**:247-287

[69] Henshall AS, McInnes IJ. A beaker grave at springwood, Kelso, Roxburghshire. Proceedings of the Society of Antiquaries of Scotland. 1968;**100**:79-85

[70] Pearce M. What this awl means: Understanding the earliest Italian metalwork. In: Ridgway D, Ridgway FRS, Pearce M, Herring E, Whitehouse RD, Wilkins JB, editors. Ancient Italy in its Mediterranean Setting. London: Accordia Research Institute Univ. of London; 2000. pp. 67-73

[71] Barrowclough DA. The secrets of the craft production of Scandinavian late neolithic flint daggers. Lithic Technology. 2004;**29**(1):75-86

[72] Strand Tanner GH. When metal met stone: Searching for traces of metal tool utilization during the production of late neolithic nordic flint daggers [thesis]. Stockholm: University of . Stockholm; 2015. Available from: http://www.diva-portal.org/smash/get/diva2:839707/FULLTEXT01 [Accessed: October 6, 2021]

[73] Freudenberg M. Stone age or bronze age: Cushion stones and other stone tools used for early metalworking in Schleswig-Holstein. In: Eriksen BV, editor. Lithic Technology in Metal Using Societies: Proceedings of a UISPP Workshop, Lisbon, September 2006. Højbjerg: Jutland Archaeological Society; 2010. pp. 23-32

[74] Croft S. Lithic Residue Analysis: A Review and Guide to Techniques. Oxford: BAR Publishing; 2021

[75] Thomsen E. Flintens geologi og mineralogi. In: Eriksen BV, editor. Flintstudier. Aarhus: Aarhus Universitetsforlag; 2000. pp. 17-36

[76] Hein W, Lund M. Flinthandwerk. Ludwigshafen: Verlag Angelika Hörnig; 2017

[77] Eriksen BV. Implications of thermal pre-treatment of chert in the German Mesolithic. In: Schild R, Sulgostowska Z, editors. Man and Flint. Proceedings of the VII International Flint Symposium Warszawa-Ostrowiec Swietokrzyski, September 1995. Warsaw: Institute of Archaeology and Ethnology, Polish Academy of Sciences; 1997. pp. 325-329

[78] Eriksen BV. Colourful Lithics – the "Chaîne Opératoire" of heat treated chert artefacts in the early mesolithic of Southwest Germany. In: Kind CJ, editor. After the Ice Age Settlements, Subsistence and Social Development in the Mesolithic of Central Europe. Stuttgart: Konrad Theiss Verlag; 2006. pp. 147-153

[79] Olausson D, Larsson L. Heat treatment of Flint in the Scandinavian stone age? Papers of the Archaeological Institute University of Lund. 1982;**1981-1982**(4):5-25

[80] Cotterell B, Kamminga J, Dickson FP. The essential mechanics of conchoidal flaking. International Journal of Fracture. 1985;**29**(4):205-221

[81] Damlien H. Striking a difference? The effect of knapping techniques on blade attributes. Journal of Archaeological Science. 2015;**63**:122-135

[82] Floss H. Grundbegriffe der Artefaktmorphologie und der Bruchmechanik. In: Floss H, editor. Steinartefakte: Vom Altpaläolithikum bis in die Neuzeit. Tübingen: Kerns Verlag; 2013. pp. 117-132

[83] Floss H, Weber MJ. Schlagtechniken. In: Floss H, editor. Steinartefakte: Vom Altpaläolithikum bis in die Neuzeit. Tübingen: Kerns Verlag; 2013. pp. 133-136

[84] Cotterell B, Kamminga J. The formation of flakes. American Antiquity. 1987;**52**(4):675-708

[85] Pelegrin J. Lithic Technology in Harappan Times. In: Parpola A, editor. South Asian Archaeology 1993: Proceedings of the Twelfth International Conference of the European Association of South Asian Archaeologists Held in Helsinki University 5-9 July 1993. Helsinki: Suomalainen Tiedeakatemia; 1994. pp. 587-598

[86] Méry S, Anderson P, Inizan ML, Lechevallier M, Pelegrin J. A pottery workshop with flint tools on blades knapped with copper at Nausharo (Indus civilisation, ca. 2500 BC). Journal of Archaeological Science. 2007;**34**(7):1098-1116

[87] Hayden B, Hutchings WK. Whither the billet flake? In: Amick DS, Mauldin RP, editors. Experiments in Lithic Technology. Oxford: BAR Publishing; 1989. pp. 235-257

[88] Waldorf DC. The Art of Flintknapping. 4th ed. Branson, MO: Mound Builder Books; 1993

[89] Pelegrin J. Long blade technology in the old world: An experimental approach and some archaeological results. In: Apel J, Knutsson K, editors. Skilled Production and Social Reproduction Aspects of Traditional Stone-Tool Technologies [Symposium Uppsala, 2003]. Uppsala: Societas Archaeologica Upsaliensis; 2006. pp. 37-68

[90] Anderson-Gerfaud P, Inizan ML, Lechevallier M, Pelegrin J, Pernot M. Des lames de silex dans un atelier de potier harappéen. Interaction de domaines techniques. Comptes rendus de l'Académie des Sciences de Paris. 1989;**308**:443-449

[91] Nicolas C, Guéret C. Armorican arrowheads biographies: Production and function of an early bronze age prestige good from Brittany (France). Journal of Lithic Studies. 2014;**1**(2):101-128

[92] Guilbeau D. Les grandes lames et les lames par pression au levier du Néolithique et de l'Énéolithique en Italie [thesis]. Paris: Paris 10; 2010. Available from: https://www.theses.fr/2010PA100146 [Accessed: November 15, 2023]

[93] Semenov SA. Prehistoric Technology: An Experimental Study of the Oldest Tools and Artefacts from Traces of Manufacture and Wear. London: Cory, Adams & Mackay; 1964. 211 p

[94] Anderson PC. A testimony of prehistoric tasks: Diagnostic residues on stone tool working edges. World Archaeology. 1980;**12**(2):181-194

[95] van Gijn A. The interpretation of "sickles". A cautionary tale. In: Anderson PC, editor. Prehistory of Agriculture New Experimental and Ethnographic Approaches. Los Angles: Institute of Archaeology University of California; 1999. pp. 363-372

[96] Pawlik AF. Identification of hafting traces and residues by scanning electron microscopy and energy-dispersion analysis of X-rays. In: Walker EA, Weneban-Smith F, Healy F, editors. Lithics in Action. Oxford: Oxbow; 2004. pp. 169-179

[97] Longo L, Skakun N, Saracino M, Dalla Riva M, editors. 'Prehistoric Technology' 40 Years Later: Functional Studies and the Russian Legacy: Proceedings of the International Congress Verona (Italy) 20-23 April 2005: Proceedings of the International Congress Verona (Italy) 20-23 April 2005. Ann Arbor, MI: University of Michigan Press; 2008

[98] Pedergnana A, Ollé A. Building an experimental comparative reference collection for lithic micro-residue analysis based on a multi-analytical approach. Journal of Archaeological Method and Theory. 2018;**25**(1):117-154

[99] Freudenberg M. Steingeräte zur Metallbearbeitung. Einige Neue Aspekte zum Spätneolithischen und Frühbronzezeitlichen Metallhandwerk vor dem Hintergrund des Schleswig-Holsteinischen Fundmaterials. Archäologisches Korrespondenzblatt. 2009;**39**:341-359

Section 2

Biological Applications

Chapter 3

Novel Synthesis of N–N Azo and Hydrazine Phenyl Ligand Derivatives for Copper(II) Complex Bio-Active Application

Ahlam I. Al-Sulami and Tesfay G. Ashebr

Abstract

Copper(II) complexes possess relatively a broad spectrum of medicinal importance with less toxicity. It is important to note that, in this chapter, copper(II) is selected as chelating central metal atom considering its current reputation to design such bio-active compounds due to its; (i) permits in realizing stable coordination compound, (ii) diverse coordinating capability with oxygen (O), nitrogen (N), sulfur (S) and phosphorus (P) donor ligands, and (iii) exhibits potentially better biological activity. Therefore, the presented chapter offers the up-to-date advancement and future perspective of bio-active copper(II) complexes derived from Schiff base of azo- and phenyl hydrazine-based ligands and their derivatives. To showcase the existing trend of these classes of bio-active compounds, due to the wide depth of the literature, selected seminal compounds exhibiting outstanding biological activity are discussed in detail. Recent studies establish that azo- and phenyl hydrazine-based bio-active copper(II) complexes are among the promising candidates that are expected to replace the conventional antibiotics which are suffering from side effects as well as microbial resistance. However, the collaborative efforts of chemists and biotechnologists are still needed to realize their real world application.

Keywords: bio-active compounds, copper(II), ligands, copper(II) complexes, azo- and phenyl hydrazine ligands

1. Introduction

Medicinal inorganic chemistry deals with searching and designing of additional therapeutic alternatives not accessible to the organic compound counterparts, in which metal complexes have been extensively investigated. This is because of the wide range of coordination numbers, oxidation state, coordination geometries, and accessible redox states, which are the intrinsic properties of the Lewis acid central metal ions as well as the different coordination modes of donor atoms (N, O, S, and so on) of the Lewis base ligands offer the medicinal inorganic chemist a large opportunity to extensively exploited [1, 2]. Of the transition metals, copper plays a crucial part in biological processes in mitochondria and cell [3, 4]. For example, **Figure 1** demonstrates copper homeostasis i.e. the mode of copper uptake, distribution as well as removal in mammalian cells. Copper can also form stable redox-active biological compounds with a variety of ligands that contain donor atoms like O, S, or N [6]. Moreover, the biological accessible redox potential of copper as Cu(II)/Cu(I) redox pairs have biologically significant implications such as metallo-nucleases [7, 8]. Copper is not only necessary for numerous biological processes but also a powerful anti-microbial tool that can be used to combat invasive diseases [9]. Consequently, copper complexes have been extensively studied for their potential biological activity such as antimicrobial [10], antiviral [11], anti-inflammatory [12], and antifungal [13] properties.

Researchers have been realized that, compared to the ligands, metal complexes could possess a greater biological activity with lower toxicity [14]. Of the metal

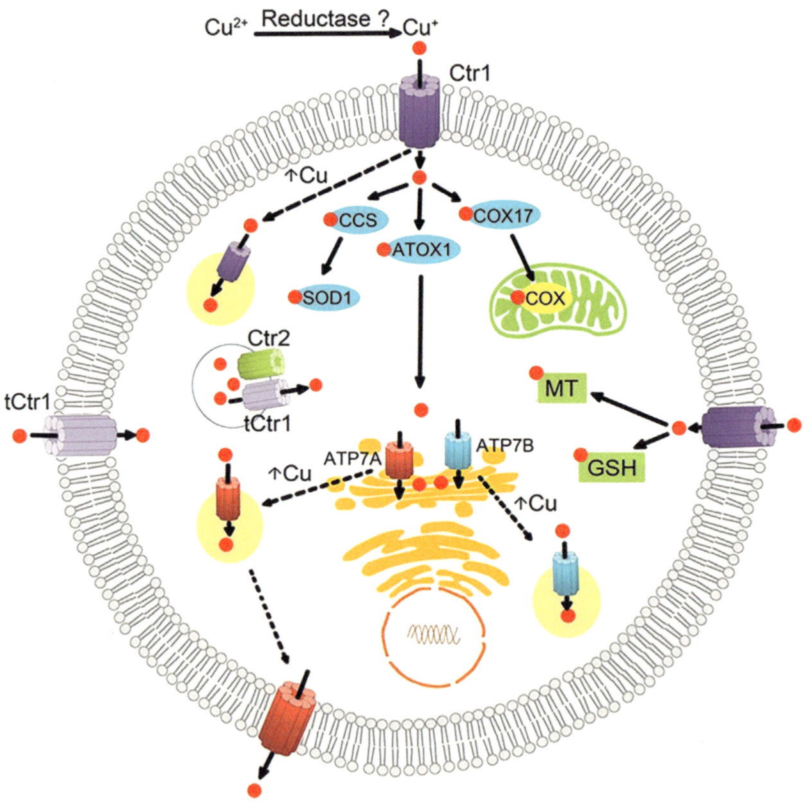

Figure 1.
Copper homeostasis and its mode of uptake, distribution as well as removal in mammalian cells. After reduction of Cu(II) to Cu(I) form, it enters the cell via the copper importer Ctr1 and then passed on to the chaperones COX17, CCS, and ATOX1, which deliver the copper to the cytosolic SOD1, COX in the mitochondria and to ATP7A/B at the TGN, respectively. Moreover, copper binds to cellular the antioxidant (MTs and GSH) that helps to prevent the formation of free catalyzing copper ROS. At the TGN, copper is amalgamated with copper-dependent enzymes (such as ceruloplasmin) that migrate through the secretion pathway. Under intracellular copper elevated condition, its excretion is facilitated via ATP7A as well as ATP7B circulation from the TGN to the plasma membrane, while the transporter Ctr1 is suppressed and is subsequently degraded. Ctr2 can also increase the generation of tCtr1, which facilitate endosomal copper transport to the cytoplasm resulting in the diminishing of the intracellular copper accretion [5]. Abbreviations: Trans-Golgi network = TGN, ATP7A/B = copper transporting ATPase A/B, ATOX1 = antioxidant protein, COX = cytochrome c oxidase, COX17 = cytochrome c oxidase copper chaperone, CCS = copper chaperone for SOD1, ROS = reactive oxygen species, Ctr1/2 = copper transporter 1/2, MT = metallothioneins, SOD1 = Cu/Zn–superoxide dismutase, GSH = glutathione, and tCtr1 = truncated Ctr1.

complexes, copper complexes mainly depend on the oxidation state i.e. copper(I) and copper(II) as well as nature of the ligands and their donor atoms [9]. Like other transition metal ions, copper has various oxidation states, i.e. I, II, and III, of which the copper(II) with a d^9 system of electronic configuration is the most favorable and relatively stable oxidation state. Considering its broad spectrum of bioactivity application, copper(II) is relatively vital in medicinal and pharmaceutical fields. Apart from the central metals, a rational designing of the ligands is also a very important step in developing bioactive complexes. For this reason, several ligands have been employed such as, Schiff bases ligands with multi dentate scaffold bearing imine nitrogen (C=N) which can be realized by simple condensation of aldehydes or ketones. Moreover, azo-based ligands possessing the azo functional group (N=N) moiety which can be also prepared by diazotization-condensation [15] as well as hydrazone-based ligands which are another class of Schiff base compounds containing hydrazine moiety (C=N–N) which can be synthesized by hydrazine-aldehyde/ketone condensation or Japp-Klingemann reaction [16, 17] have been also employed. The strategic synthesis of such ligands is well documented and extensively explored. Therefore, due to their straightforward synthesis route of these classes of ligands, their detailed synthesis strategy is omitted to avoid redundancy and only the type of ligands covered in this chapter.

Copper(II) complexes can display a substantial biological activity employing different ligands such as Schiff bases containing azo-, hydrazine-, and hetrocycles-moiety. These types of ligands containing imine moiety can form a stable metal coordination complexes which are responsible for broad spectrum of applications of which biological activity is one application field [18]. The presented chapter offers the current advancement and forwards the future perspective of copper(II) complexes of both azo- and hydrazine-based ligands and their biological importance. Appreciating all efforts that have been done so far, due to the wide depth of the literature, only selected seminal reports of azo- as well as phenyl hydrazine-based copper(II) complexes exhibiting superior biological activity are considered in this chapter. Additionally, more emphasis is also given to the complexes which are structurally well characterized (using single crystal XRD) unless the complexes reported with powder chemistry structural exhibited outstanding biological activity. Finally, the theoretical consideration of some selected seminal works is also incorporated to probe their relationship with the experimental aspect.

2. Bio-active copper(II) compounds

The bio-activity, biocompatibility as well as versatile coordination chemistry nature of the copper metal offer an opportunity for realizing bio-active copper(II) complexes which is usually achieved via enhancing the initial bio-activity of the ligand such as Azo-Schiff base ligands [15] and hydrazone based ligands [19]. Incorporating two or more diverse chromophores into the ligand could potentially enhance the physicochemical as well as the biological properties of the desired ligands and their respective metal complexes [15]. Azo are typically Schiff base ligands possessing a distinct chromophore group (N=N) which is responsible for diverse potential range of applications such as dyes [20] and biological [21] applications. Moreover, hydrazine and/or hydrazine based ligands displaying keto-enol tautomerism are another class of Schiff base ligands where the complexation of copper(II) with such biologically active ligands offers the opportunity of improving the efficacy of anti-microbial activity via evading the drug resistant pathogens [19, 22]. In this chapter, the biological activity of various N–N azo- as well as phenyl hydrazine ligand and/or hydrazone derivatives for copper (II) complex is summarized, including some copper(II) complexes based drugs commonly used as antibiotics, antimicrobials, and anti-inflammatory remedies.

2.1 Azo-based bio-active copper(II) complexes

Apart from industrial applications, dyes mainly azo-dyes possess massive biological applications (**Figure 2**) confirming the versatility of these class of compounds. Recently, the biological application of azo-based ligands and their metal complexes are getting much attention due to the presence of the optically and biologically active azo- functional group. Azo compounds are commonly prepared by diazotization and coupling i.e. diazotization of an aromatic primary amine followed by coupling with electron-rich nucleophile such as amino and hydroxyl group under alkaline conditions. Due to straight forward synthesis of the azo-based multi dentate ligands and their readily coordination flexibility, azo-based copper(II) bio-active complexes has been extensively explored [15, 21, 23]. For example, more recently Mady et al. reported a series of o-vanillin-azo (OV-AZO) based ligand prepared by condensation of 4-aminoazobenzene and o-vanillin and their transition metal complexes (Mn(II), Ni(II), Co(II), Cu(II), Zn(II), and Zr(IV)) for their anticancer and antimicrobial

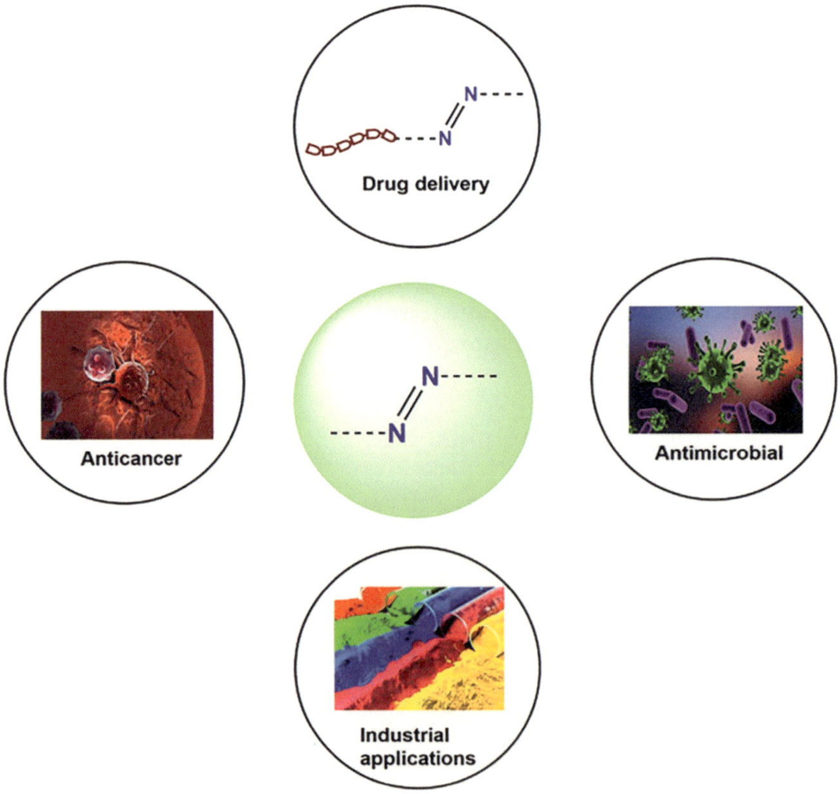

Figure 2.
Schematic illustration of azo-dyes and their biological as well as industrial applications.

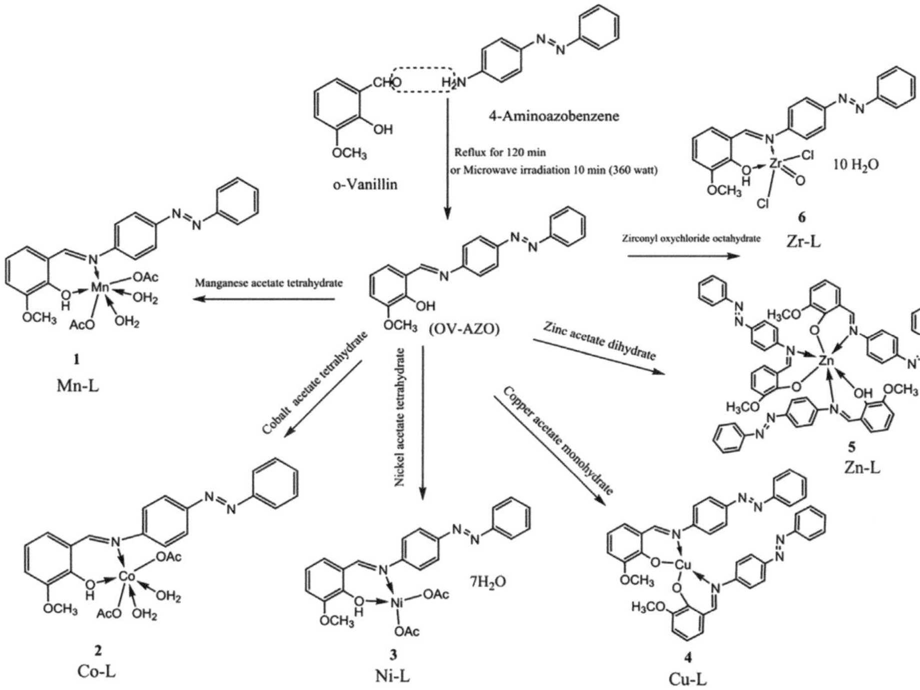

Figure 3.
Proposed structures of the azo-based ligand (OV-AZO) and its transition metal complexes [23].

activity (**Figure 3**) [23]. The obtained results showed that, in comparison to the free corresponding azo-based ligand, of the other transition metal complexes, the copper(II) complex exhibited the maximum cytotoxic activity of an half maximal inhibitory concentration (IC_{50}) of 18 and 22 μg/mL for human colon and liver cancer cells, respectively as compared to the reference cisplatin drug. This could be due to the copper(II) complex capability to induce oxidative stress as well as generating activated oxygen species in addition to its structural arrangement offering relatively better stability advantage while the other counterparts were suffering from water dehydration and acetate removal. Moreover, similarly, higher antimicrobial activity of metal complexes against the anti-microbial (fungal and bacterial strains) was witnessed as compared to the free corresponding ligand which could be due to the Tweedy's chelation theory. Normally, IC_{50} is a quantitative measure of the effectiveness of a compound in inhibiting a specific biochemical or biological function that can be determined by constructing a dose–response curve assay at different concentrations. Usually, IC_{50} is typically expressed as molar concentration where in this chapter a molar concentration in μg/mL is used as a figure of merit.

Kurtoglu and coworkers reported azo-azomethine based ligands of Cu(II) complexes and their antioxidant and cytotoxicity against human uterus carcinoma cells lines (**Figure 4**) [24]. Interestingly, all the ligands and the complexes exhibited better antiproliferative activities at 25, 50, and 100 $\mu g mL^{-1}$ concentrations where complexation of the ligands with the copper(II) ion slightly declined their anticancer properties which could be due to the hindering of the imine, phenolic and methoxy groups. Similarly, the antioxidant properties of the ligands were slightly better than the copper(II) complexes. Moreover, Krishnamurthy and coworkers reported an

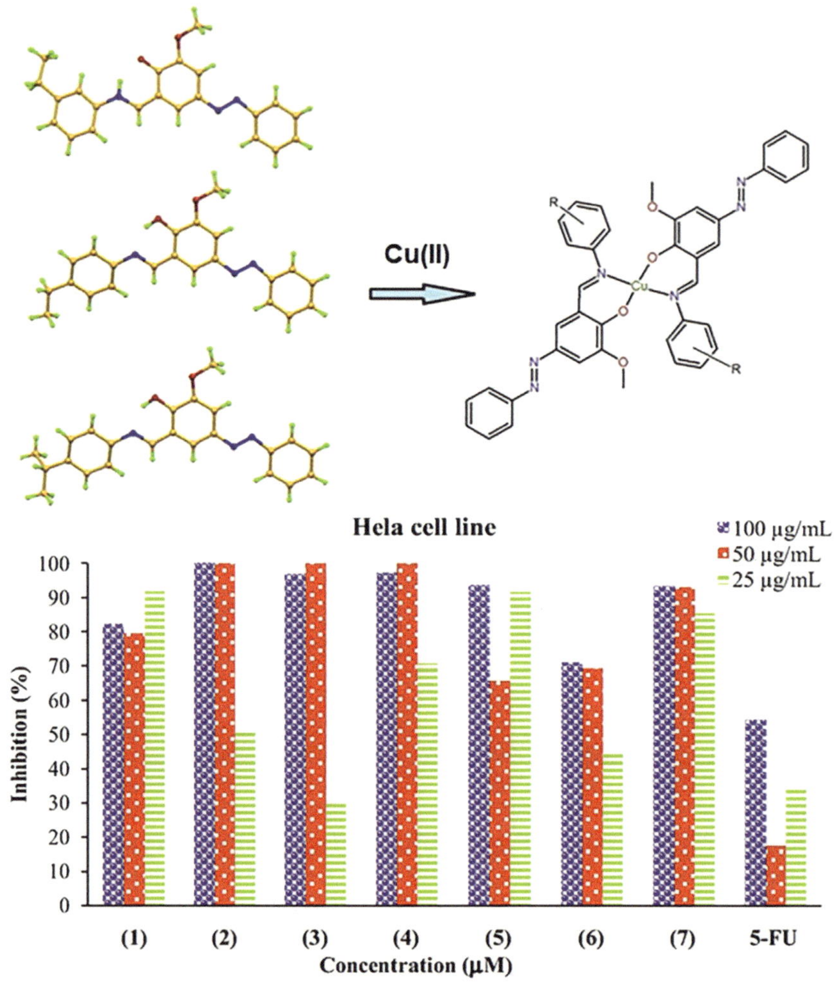

Figure 4.
Crystal structure of the azo-azomethine based ligands, their Cu(II) complexes and their antiproliferative activities where 1 is the azo-aldehyde, 2-4 are the ligands (R = 3-ethylaniline or 3-CH_2CH_3; 4-ethylaniline or 4-CH_2CH_3, and 4-isopropylaniline or 4-$CH(CH_3)_2$), 5-7 are the complexes and 5-Fu is fluorouracil [24].

isonicotine hydrazide based azo- ligand and its transition (Cu(II), Co(II), and Ni(II)) metal complexes with octahydral geometry displaying biological activity such as, human breast cancer cells (MDA-MB 231) where MDA stands for M.D. Anderson and MB stands for mammalian breast (**Figure 5**), as well as deoxyribonucleic acid (DNA) cleavage activity [25]. Scanning electron microscope (SEM) images revealed that the presence of variation in ligand and metal complexes surface morphology which could be because of the change of metal ion i.e. broken rock like, and ice block like structure for ligand and copper(II) complex, respectively as well as Cotton-like filamentary morphology for both cobalt(II) and nickel(II) complexes (**Figure 5**, bottom). Of the transition metal complexes, copper(II) complex exhibited the highest biological activity against all the investigated cell lines confirming that biological activities depends on the type of the central metal ion.

Keshavayy and coworkers reported benzothiazole based azo- ligand with indol wing and its transition metal complexes (Cu(II), and Ni(II) with distorted square

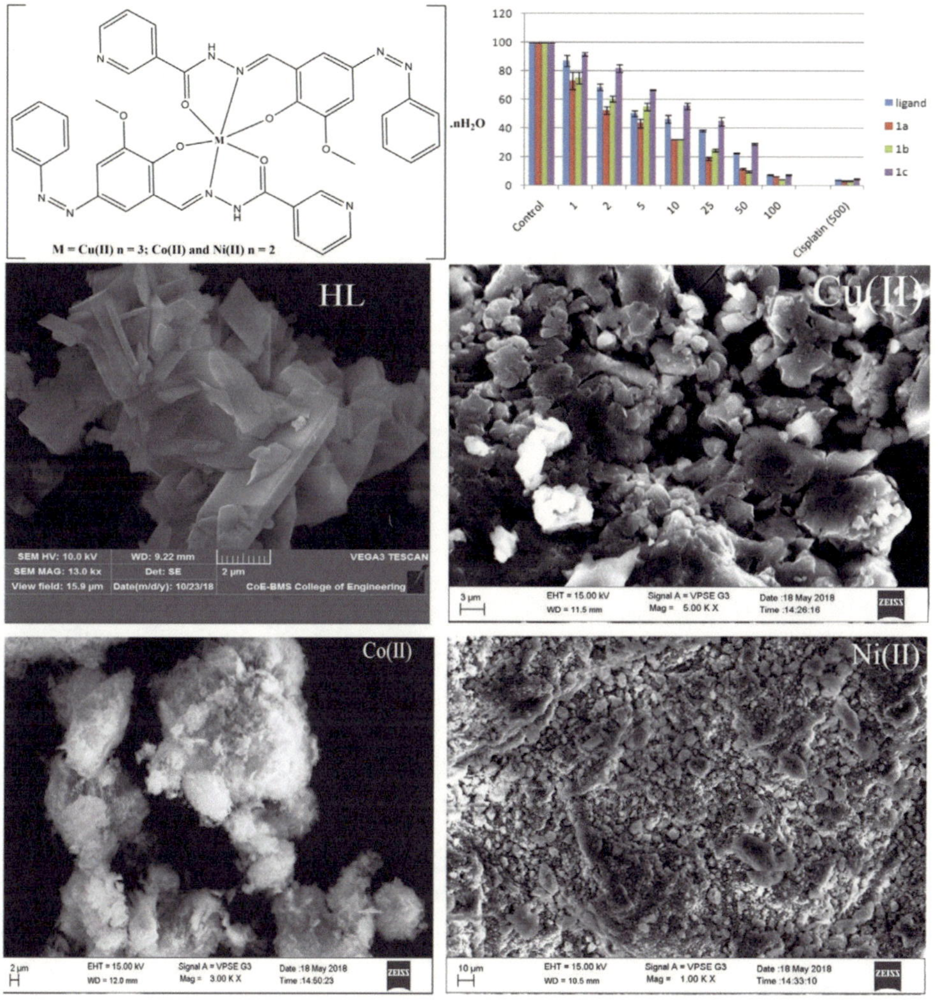

Figure 5.
Schematic representation of the metal complexes and their cytotoxicity effect by methyl thiazolyl tetrazolium (MTT) assay of ligand as well as the complexes on breast cancer cells (MDA-MB-231) cells (top) as well as their SEM morphological structure of the ligand (HL) and the corresponding metal complexes (bottom) [25].

planar, Co(II), and Fe(III) with octahedral as well as Zn(II) with tetrahedral geometry) for their potential antimicrobial, antitubercular as well as DNA cleavage activities (**Figure 6**) [26]. Apart from the investigated biological activities, all the formulated compounds demonstrated a promising anti-tubercular activity i.e. against *M. tuberculosis* (**Figure 6**, right). Moreover, in their extended work, i.e. benzothiazole based azo- ligand possessing a pyrazole ring and its transition metal complexes (Cu(II) with distorted tetrahedral, Ni(II), and Co(II) with octahedral geometry) for their anti-bacterial, anticancer and DNA cleavage activity (**Figure 7**) [27]. Apart from the promising anti-bacterial, and anticancer properties, copper(II) and nickel(II) complexes exhibited complete cleavage of all forms of DNA which could be due to the effective coordination role of oxygen and nitrogen, while ligand as well as cobalt(II) complex displayed partial DNA cleavage compared to the control DNA. Moreover, Keshavayy and coworkers also reported another benzothiazole based azo- ligand with

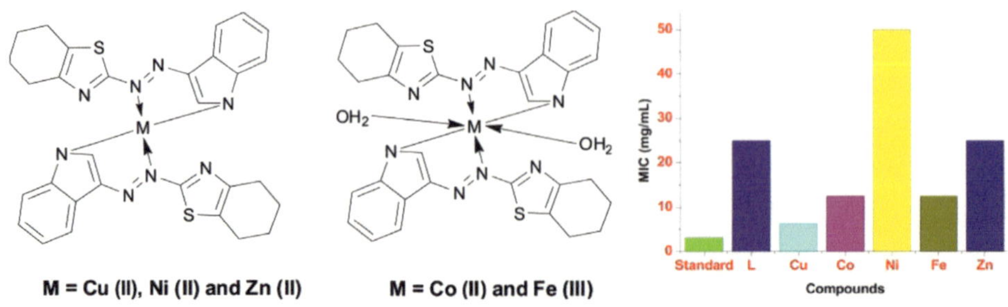

Figure 6.
Proposed schematic structures of the metal complexes and the anti-tubercular activity of the ligand and its metal complexes [26].

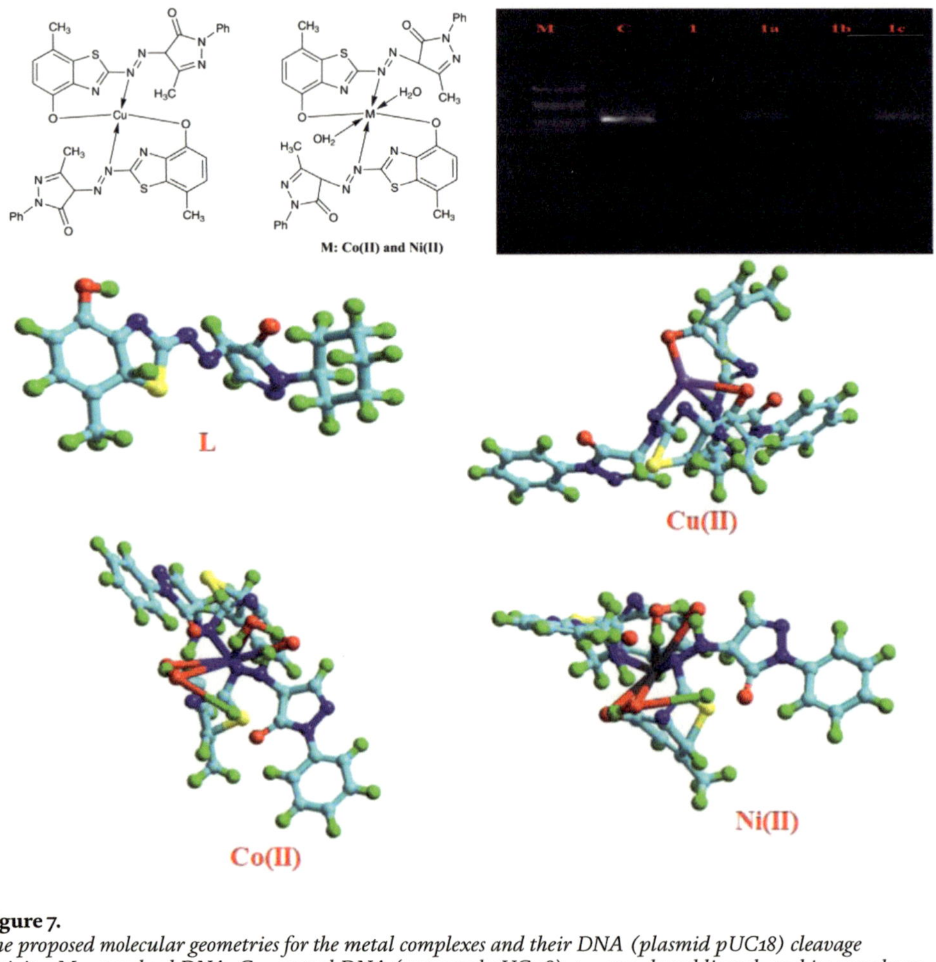

Figure 7.
The proposed molecular geometries for the metal complexes and their DNA (plasmid pUC18) cleavage activity; M = standard DNA, C = control DNA (untreated pUC 18), 1 = azo- based ligand, and its complexes (1a = Cu(II), 1b = Co(II) and 1c = Ni(II), respectively) (top). The molecular geometries (calculated) of the ligand (L) as well as its complexes (bottom) [27].

pyridone moiety and its transition metal complexes (Cu(II), Ni(II), Co(II), Cd(II), and Zn(II)) for their potential biological activity (**Figure 8**) [28]. All the synthesized compounds (the azo- ligand and its respective metal complexes) displayed significant

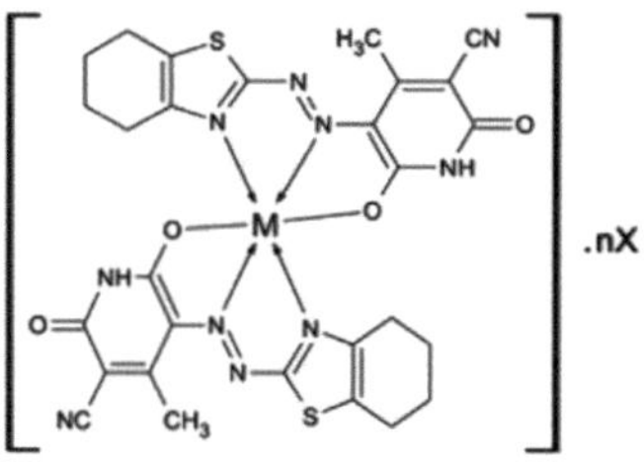

M= Co(II),n=2,X=H $_2$O
M= Ni(II),n=0,X=0
M=Cu(II), n=0, x=0

Compounds	IC_{50}		
	1 µg/mL	2 µg/mL	3 µg/mL
HL	30	35	60
$[Cu(L)_2]$	21	27	56
$[Co(L)_2(H_2O)_2]$	19	24	42
$[Ni(L)_2]$	17	26	42
$[Zn(L)_2]$	20	22	51
$[Cd(L)_2]$	15	26	35
Ascorbic acid	10	11	20

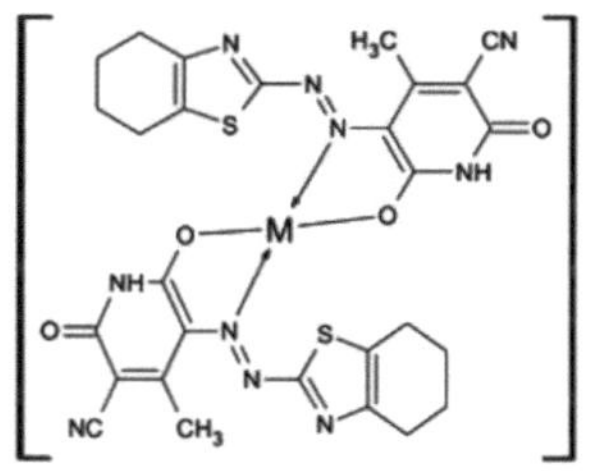

M= Zn(II), Cd(II)

Compounds	IC_{50} (µg/mL)		
	Jurkat	A549	K562
HL	53.59	38.10	53.12
$[Cu(L)_2]$	23.16	24.13	50.06
$[Co(L)_2(H_2O)_2]$	27.71	42.16	53.12
$[Ni(L)_2]$	42.41	23.31	14.52
$[Zn(L)_2]$	15.36	32.25	50.31
$[Cd(L)_2]$	10.12	35.12	13.25

Figure 8.
The proposed molecular geometries for the metal complexes and their biological activity [28].

DPPH (2,2-diphenyl-1-picrylhydrazyl) radical scavenging potential compared with the standard ascorbic acid as well as good anticancer activity against Jurkat, K562, and A549 cell lines (**Figure 8**).

More recently, copper(II) and cobalt(II) complexes of azo- ligand obtained via the diazotization reaction of 2-amino pyridine with pyrazolone and their DNA cleavage and antibacterial (Klebsiella pneumonia, and *Bacillus subtills*) activity was reported by Channabasappa and coworkers (**Figure 9**) [29]. The show that Copper(II) and Cobalt(II) complexes exhibited promising DNA Cleavage activity as well as antibacterial against *K. pneumonia* and *B. subtills* as compared to the standard ampicillin properties. Moreover, the azo-ligand and its respective metal complexes displayed well-established bonds with the target amino acids (**Figure 9**, right).

More recently, our group also reported an azo- ligands possessing pyridone moiety with long aliphatic chain and their respective nitro and methyl group substituent as well as solvent dependent copper(II) complexes (elongated octahedral geometry) targeting its potential inhibitory against Alzheimer's disease in comparison with Donepezil drug [21]. Our findings revealed that the binding affinity of the ligand as well as its copper(II) complexes were found with 1EVE ranges from −9.6 to −8.3 kcal/mol while −10.9 kcal/mol is for Donepezil drug suggesting our complexes could be promising drug candidates for Alzheimer's disease. Moreover, comparing to both complexes, the Donepezil drug and ligands were found functioning as carcinogenic.

Recently, in continuation of our work on azo- ligands owning pyridone moiety, we report a diazenyl pyridinone based heterocyclic ligands and their Cu(II) complexes (octahedral geometry) for their potential antimicrobial, anticancer, and antioxidant properties (**Figure 10**) [30]. Our results revealed that the complexes showed better antitumor activity as well as selectivity index against the human primary hepatocytes THLE-2 and breast adenocarcinoma cells MCF-7 than the reference doxorubicin.

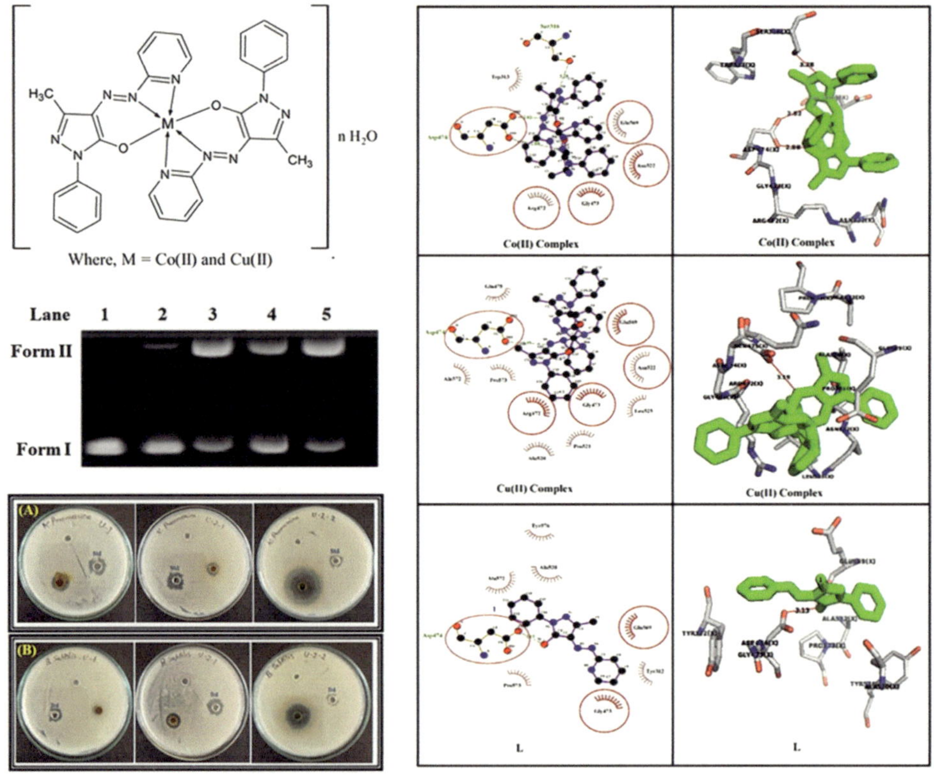

Figure 9.
The proposed molecular geometries for the metal complexes, DNA cleavage of by the Cu(II), and Co(II) complexes at body temperature where lane 1 DNA alone; lane 2 = DNA with 40 μl of copper(II) complex; lane 3 = DNA with 60 μl of copper(II) complex; lane 4 = DNA with 40 μl of cobalt(II) complex; lane 5 = DNA with 60 μl of cobalt(II) complex; form I = supercoiled and form II = nicked circular DNA, respectively as well as zone of inhibition of the azo- ligand and their metal complexes against (A) Klebsiella pneumonia and (B) bacillus subtills; where u-1 = azo- ligand, u-2-1 = Co(II), and u-2-2 = Cu(II) complexes, respectively (left). In silico molecular docking investigations of the azo- ligand as well as its respective metal complexes (right) [29].

Moreover, the prepared compounds exhibited significant anti-bacterial and anti-fungal properties (**Figure 11**, right).

More interestingly, the azo- pyridone moiety containing ligands and their copper(II) complexes are becoming biologically promising where Lađarević and coworkers also reported on a similar work by ligand substituent tuning strategy [31]. In this work, both complexes, the chloride substituted (Cu(II)-Cl) and methoxy substituted (Cu(II)-OCH_3) was found to be promising compounds for the treatment of various types of cancer which could be considered as potential therapeutic candidates. Among the two complexes, The most favorable structure exhibiting the best docking potential toward both proteins was the complex Cu(II)-OCH_3 bearing methoxy group. Moreover, the Cu(II)-OCH_3 complexes exhibited the highest antioxidative activity compared with its parent ligand and the standard ascorbic acid antioxidant which could be due to the presence of an electron–donor methoxy group making this complex as a favorable antioxidant agent. This findings revealed that both anticancer and antioxidant activities exhibited substituent dependent biological activity where for the azo- pyridone moiety containing

DOI: http://dx.doi.org/10.5772/intechopen.1004323

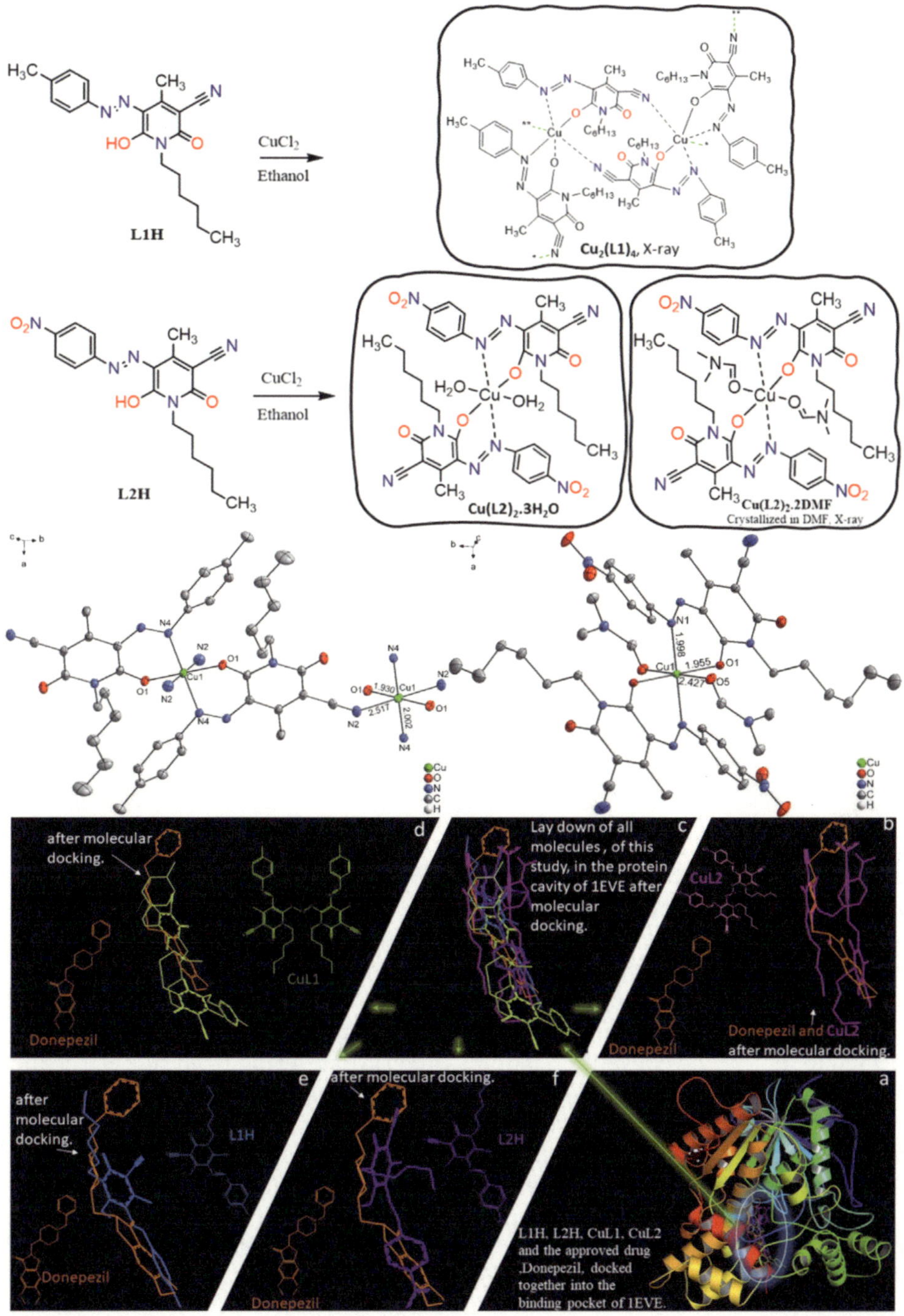

Figure 10.
Schematic illustration of the synthetic route of the copper(II) complexes employing the azo- ligand (top); the molecular structure of complexes of CuL1 (CuO_2N_4-octahedron) and CuL2 (CuO_4N_2-octahedron) (middle); as well as (a) simultaneous docking into the binding pocket of the 1EVE under similar conditions for the the superpositions of azo- ligands (L1H and L2H), complexes (CuL1 and CuL2) and donepezil drug (orange color) appeared in each window in comparison to the appearance of each ligand, (b) CuL2 and donepezil; (c) all molecules and donepezil; (b) CuL2 and donepezil; (d) CuL1 and donepezil; (e) L1H and donepezil; and (f) L2H and donepezil [21].

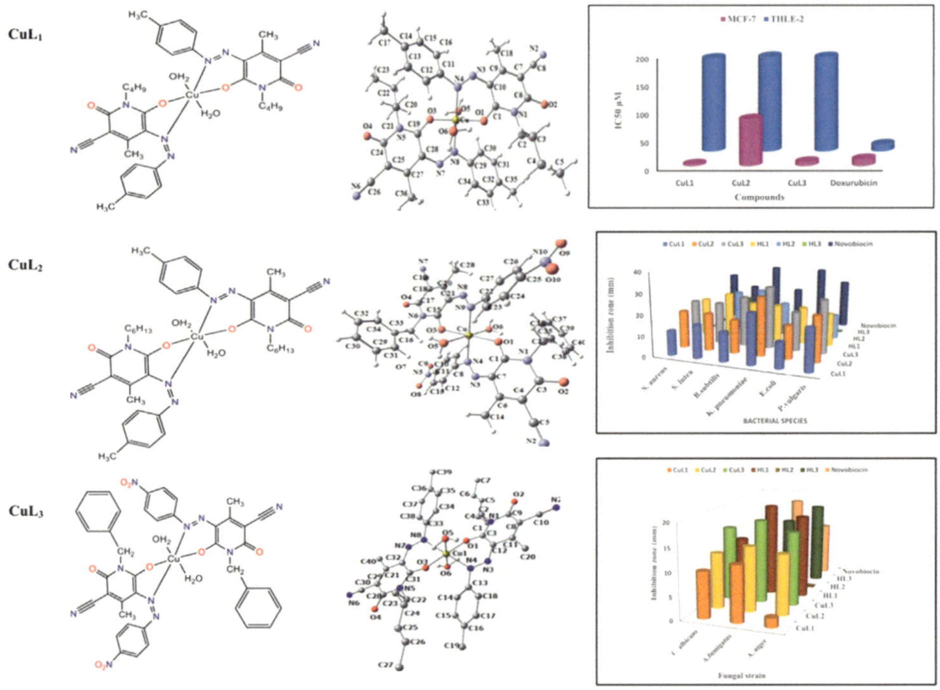

Figure 11.
Schematic illustration and the optimized molecular structure of the copper(II) complexes (left); cell viability results of copper(II) complexes against the human breast adenocarcinoma MCF-7 cells at different (0–100 μM) concentrations, anti-bacterial and anti-fungal activities [30].

ligands an electron donating methoxy group was found favorable for designing copper(II) bio-active compound (**Figure 12**).

It have been witnessed that azo- based ligands (derived from o-vanillin, salicylic, isonicotinic, thiazole, indole, pyrazole, pyridinone, and their derivatives, for details see Section 2.1) and their copper(II) complexes are found to be a favorable candidates for realizing potentially promising bio-active compounds. Such compounds are expected to replace the conventional drags which are suffering from side effect and biological pathogens resistance. Due to the vast depth of the literature with regard to azo- based copper(II) bio-active complexes, we attempt to discuss some selected promising compounds in the above. Therefore, **Table 1** below attempts to incorporate some other works which are discussed in the above in detail. Thus, we encourage readers to cross refer the compounds presented in **Table 1** for further detailed structural insights as well as to refer recent comprehensive review detailing Azo- based transition metal complexes and their biological activity reported by Younas and coworkers [15].

2.2 Phenyl hydrazine- and hydrazone-based bio-active copper(II) complexes

Hydrozones are very essential intermediates for the synthesis of different heterocyclic compounds. Acid hydrazides as well as their derivatives are also another class of hydrazine based compounds which useful chemical synthons for synthesis of

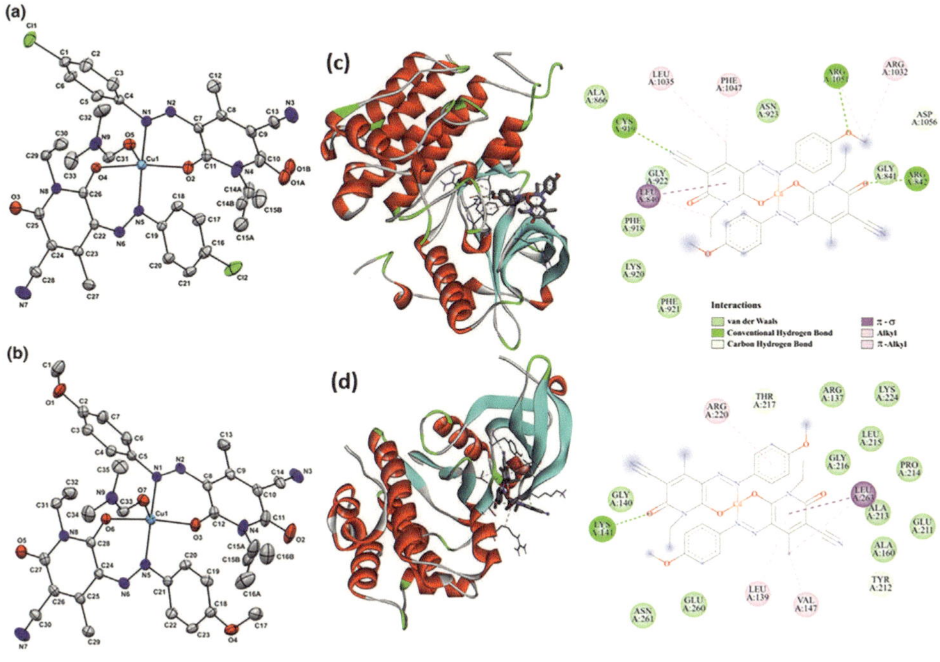

Figure 12.
Optimized molecular structure of the copper(II) complexes of Cu(II)-Cl (a) and Cu(II)-OCH_3 (b). Best docking poses with interacting residues, as well as intermolecular interactions of CuL-OCH_3complex with VEGFR-2 (c) and Aurora kinase A (d) [31].

numerous heterocyclic compounds that with pronounced biological as well as pharmacological applications [40]. Apart from biological applications, Brady and Elsmie reported that, of the nitro-substituted phenylhydrazine, dinitrophenylhydrazine is a reddish yellow compound which has been used as an analytical reagent for the identification of carbonyl functional groups quantitatively [41] e.g. in biological samples [42]. This confirms that phenylhydrazine and their derivatives are good nucleophile for carbonyl ketones as well as aldehydes in the formation multi-functional hydrazone based ligands (**Figure 13**) and their metal complexes for different applications such as biological activity [16, 17]. Currently, hydrazone ligands possessing oxygen and nitrogen are under extensive exploration due to their ease of synthesis, chelating ability, potential biological as well as pharmacological applications [19, 40]. Amid the different metal ions, the divalent Cu(II), Ni(II), Co(II), and Zn(II) are particularly favored as they play a dynamic role in biological activities [43, 44]. Due to the vast reports available in the literature, below only some selected phenyl hydrazine as well as hydrazone based copper(II) exhibiting potentially better biological activity are discussed in detail.

Idemudia and coworkers reported air-stable keto imine form of acylpyrazolone phenylhydrazones based ligands prepared by condensation of 4-acetyl and 4-benzoyl pyrazolone with dinitrophenylhydrazine as well as phenylhydrazine namely; benzoylphenylhydrazone (Bmpp-Ph), acetyldinitrophenylhydrazone (Ampp-Dh), as well as benzoyldinitrophenylhydrazone (Bmpp-Dh) and their transition metal complexes (Cu(II), Mn(II), Co(II), and Ni(II)) (**Figure 14**) [43]. All the complexes are under octahedral geometry and exhibited better biological activity than their respective

Chemical formula*	Geometry	Characterization techniques	Biological activity	Refs.
$[Cu(L_{OV\text{-}AZO})2]$	Square planar	FTIR, UV–Vis, TGA	Antibacterial	[32]
$[Cu(L1_{OV\text{-}AZO})].3H_2O$	Square planar	UV–Vis, FTIR, XRD	Anticancer	[33]
$[Cu(L2_{OV\text{-}AZO})Cl].2H_2O$	Square planar	FTIR, UV–Vis, TGA, EPR, XRD	Antibacterial and Antifungal	[34]
$[Cu(L_{SAL\text{-}AZO})].H_2O$	Square planar (distorted)	UV–Vis, FTIR, MS	Molluscicidl	[35]
$[Cu(L1_{SAL\text{-}AZO})H_2O]$	Square planar (distorted)	FTIR, UV–Vis, MS, ESR, XRD	Antibacterial and Antifungal	[36]
$[Cu(L2_{SAL\text{-}AZO})\ H_2O](OAc)_2$	Tetrahedral/ Square planar	FTIR, UV–Vis, TGA	Antibacterial	[37]
$[Cu(L3_{SAL\text{-}AZO})_2(H_2O)_2]$	Octahedral	UV–Vis, FTIR, ESI-MS	Antibacterial and Antifungal	[38]
$[Cu(L_{TSC\text{-}AZO})H_2O]$		FTIR, UV–Vis, TGA	Antibacterial	[39]
$[Cu(L_{PYRI\text{-}AZO})DMF]$	Square-pyramidal	ICP-OES, UV–Vis, FTIR, TGA, Single crystal X-ray	Antioxidant, and Anticancer	[31]
$[Cu(L_{PYRA\text{-}AZO})].nH_2O$	Octahedral	UV-Visible, FT-IR, NMR, MS, XRD	DNA cleavage and Antibacterial	[29]
$[Cu(L_{THAI\text{-}AZO})]$	Octahedral	FTIR, UV–Vis, TGA, ESR, XRD, MS, EDAX, SEM	Anticancer	[28]
$[Cu(L_{INDO\text{-}AZO})]$	Square planar (distorted)	FTIR, UV–Vis, MS, ESR, XRD	DNA cleavage and Antibacterial	[26]
$[Cu(L_{NICO\text{-}AZO})].3H_2O$	Octahedral	FTIR, UV–Vis, MS, ESR, XRD	DNA cleavage and Cytotoxicity	[25]

**The abbreviated expressions in subscript position; OV, SAL, TSC, PYRI, PYRA, THAI, INDO, and NICO stands for o-vanillin, salicylic, thiosemicarbazide, pyridinone, pyrazole, thiazole, indole, nicotinic, and their derivatives respectively, for more structural details of the ligands, please refer the original article from the provided references.*

Table 1.
Copper(II) complexes of azo- based ligands and their biological activities.

ligands. Moreover in their extended work on salicylic hydrazide based transition metal complexes (Cu(II), Ni(II), Co(II), and Zn(II)) in a 1:1 molar ratio and their biological activities, where both ligand substituent and central metal atom dependent biological activity in which the efficacy of the metal complexes (Cu(II) > Zn(II) > Co (II) > Ni(II)) was higher than to that of the ligands (**Figure 15**) [44].

Raman and coworkers reported an isatin-based hydrazone ligand prepared by condensation of 2,3,5-trichlorobenzaldehyde with isatin monohydrazone and their transition metal complexes (Cu(II), Ni(II), Co(II), and Zn(II)) in a 1:2 molar ratio for their potential biological activity (**Figure 16**) [45]. The results showed that all complexes exhibited effective oxidative DNA cleavage as well as promising antimicrobial activity where particularly copper(II) and zinc(II) complexes showed higher antimicrobial activity than the nickel(II) and cobalt(II) complexes.

Lodyga-Chruscinska and coworkers reported a good chelating hesperetin based hydrazine ligand (HHSB) and its copper(II) complex exhibiting promising cytotoxic activity against cancer cells (normal (LLC-PK1) cell lines and tumor (HeLa and

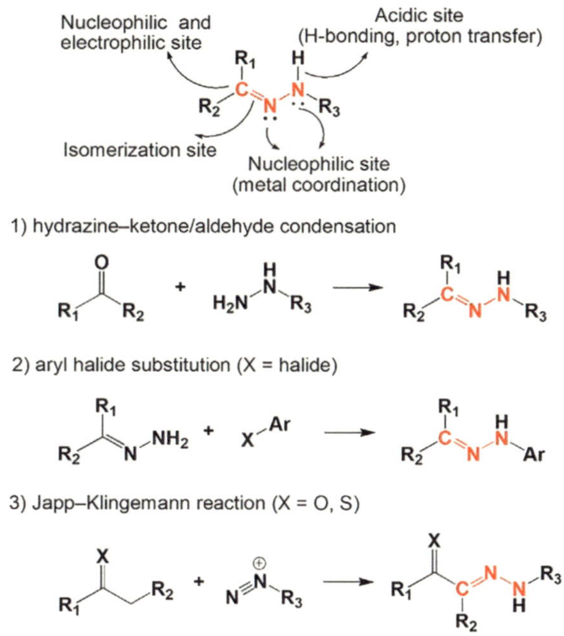

Figure 13.
Multi-functionality of hydrazone (red color) and its general synthesis route [17].

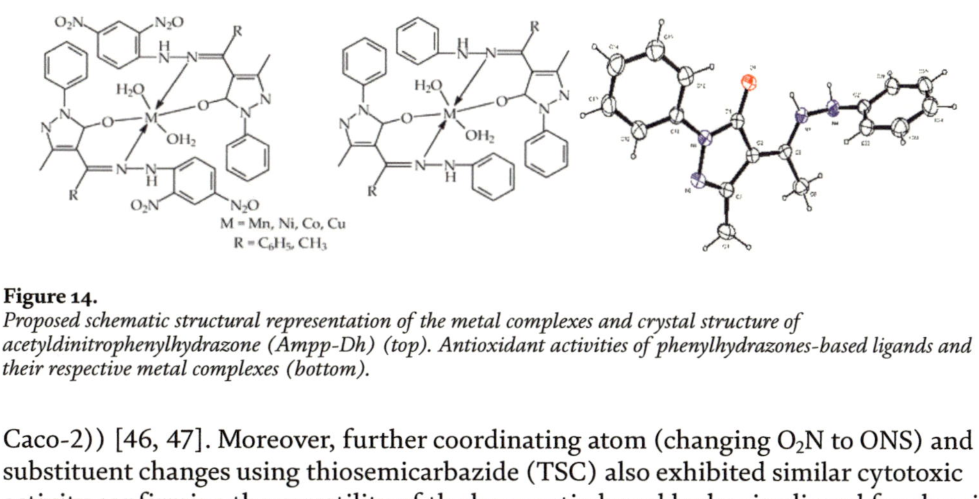

Figure 14.
Proposed schematic structural representation of the metal complexes and crystal structure of acetyldinitrophenylhydrazone (Ampp-Dh) (top). Antioxidant activities of phenylhydrazones-based ligands and their respective metal complexes (bottom).

Caco-2)) [46, 47]. Moreover, further coordinating atom (changing O_2N to ONS) and substituent changes using thiosemicarbazide (TSC) also exhibited similar cytotoxic activity confirming the versatility of the hesperetin based hydrazine ligand for chemical substituent modification (**Figure 17**).

Gatto and coworkers described pyridoxal-thiosemicarbazone and pyridoxal-S-allyldithiocarbazate hydrazone hydrazone ligands bearing ONS-donor atoms and their copper(II) complexes demonstrating promising potential anticancer activity (**Figure 18**) [48]. Interestingly, copper(II) complexes cytotoxic tests showed effective growth inhibition of both S-180 as well as Ehrlich tumor cells where enhancement of the cytotoxicity was also observed up on complexation of the ligands with copper(II) metal ions. Moreover, copper(II) complex (compound 5) demonstrated the death of

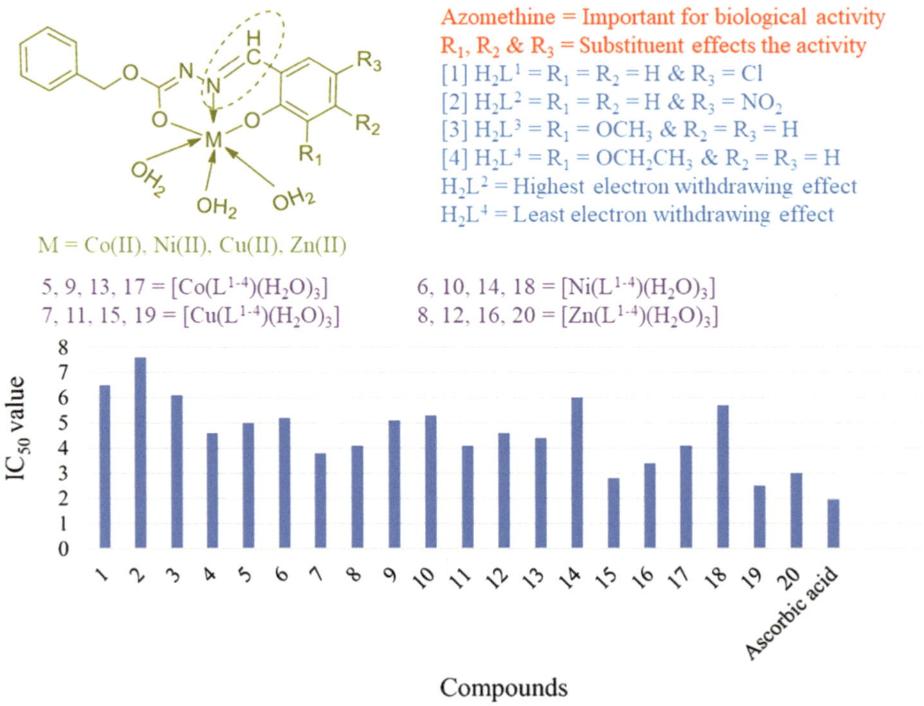

Figure 15.
Structure (proposed) and activity relationship of the complexes (top). Graphical illustration of IC_{50} value of all the compounds (bottom).

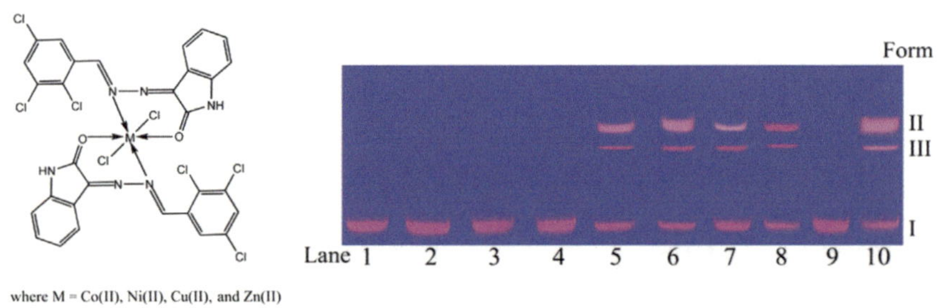

	Antibacterial activity				Antifungal activity			
Complex	*S. aureus*	*B. subtilis*	*E. coli*	*P. aeruginosa*	*A. niger*	*R. bataicola*	*R. stolonifer*	*C. albicans*
Ligand (L)	150	250	100	100	150	200	200	100
$[CoL_2Cl_2]$	75	100	25	50	50	50	75	25
$[NiL_2Cl_2]$	75	75	50	50	75	75	100	50
$[CuL_2Cl_2]$	50	50	20	25	25	20	50	20
$[ZnL_2Cl_2]$	75	100	25	20	20	25	25	20
Streptomycin	5	10	5	10	–	–	–	–
Nystatin	–	–	–	–	5	10	10	10

Figure 16.
Proposed structure and SC pBR322 DNA (0.3 μg) cleavage of by metal(II) complexes (0.30 mmol L^{-1}) in the presence of H_2O_2 (100 μmol L^{-1}) at pH 7.2. DNA control; DNA and H_2O_2; DNA + $[CuL_2Cl_2]$; DNA + ethanol (4 μL); are lanes 1–4 respectively; DNA + H_2O_2 + $[CoL_2Cl_2]$, $[CuL_2Cl_2]$, $[ZnL_2Cl_2]$, and $[NiL_2Cl_2]$, are lanes 5–8, respectively; DNA + H_2O_2 + ethanol + $[CuL_2Cl_2]$; DNA + H_2O_2 + SOD + $[CuL_2Cl_2]$ are lane 9 and lane 10 respectively (top). Antibacterial and antifungal activities of ligands, complexes and control drugs (bottom) [45].

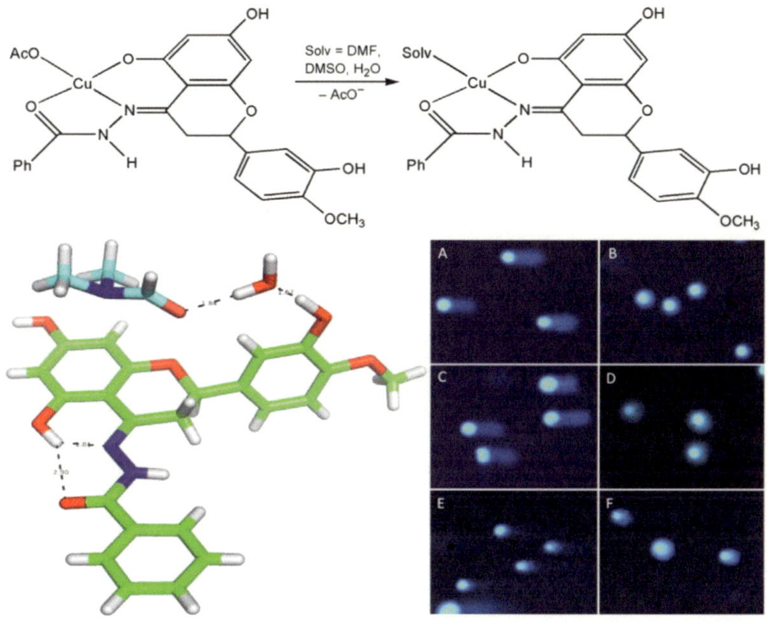

Figure 17.
Chelating ability of hesperetin based hydrazine ligand (HHSB) and its copper(II) complex proposed structure (top) [46]. Crystal structure of the hesperetin based hydrazine ligand with hydrogen bonding [46] and DNA damage of 4′,6-diamidino-2-phenylindole (DAPI) stained comets in the HeLa cells: (A) 50 μM cisplatin; (B) negative control; (C) 50 μM HHSB in Caco-2 cells; (D) 10 μM CuHHSB in LLC-PK1 cells; (E) 50 μM CuHHSB, and (F) 50 μM HHSB (bottom) [47].

the tumor cells confirming that pyridoxal derivative copper(II) complexes could be considered as great potential candidates for designing antitumor drugs which could be due to the presence of thiosemicarbazide pendant.

Naik and coworkers also reported bi-dentate indol based hydrazone ligand prepared by condensation of nicotinic acid hydrazide and indole-3-carboxaldehyde and their transition metal complexes (Cu(II), Co(II), Mn(II), and Ni(II)) under octahedral geometry for their potential biological activity (**Figure 19**) [49]. Of the transition metal complexes, copper(II) and cobalt (II) complexes exhibited better antibacterial as well as effective oxidative DNA cleavage activity. It is worthy to mention the importance of indol and nicotinic based hydrazone based ligands and their complexes suitability for designing bio-active compounds. Moreover, a similar trend where copper(II) and cobalt (II) complexes exhibited better antibacterial as well as effective DNA cleavage activity reported by Velladurai and coworkers using triazine based tridentate ligand prepared from 1,2,4-triazine and 1,4-Naphthoquinon pendants (**Figure 20**) [50].

Adimule and coworkers also reported benzimidazol based hydrazone bi-dentate ligand bearing NS donor atoms and its transition metal complexes (Cu(II), Ni(II), and Co(II)) for their DNA binding and antibacterial activity (**Figure 21**) [51]. In addition to antibacterial activity of all the complexes, among the transition metal complexes, copper (II) complex showed that an intercalative DNA binding and DNA cleavage in presence as well as absence of oxidant H_2O_2. was also observed.

Moreover, a mixed-ligand copper(II) complexes of hydrazone-based ligands displaying efficient biological activities were also reported by Annaraj and coworkers

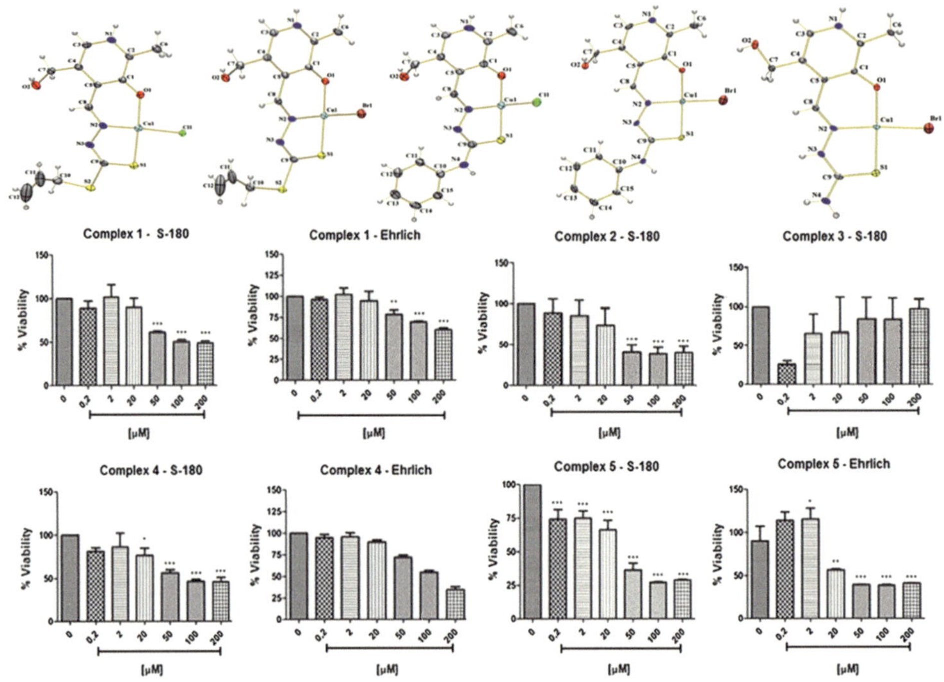

Figure 18.
*Effect of each copper (II) complex (1–5 with coordinated atoms color code: Blue N, red O, gold S, green Cl, maroon Br and cyan Cu) on the viability of the murine S-180 and Ehrlich cells treated with different concentrations with statistical p-values (*p < 0.05; **p < 0.01; ***p < 0.001) (bottom) [48].*

where the vital role of co-ligands on cytotoxicity and DNA/protein interactions has been described in detail (**Figure 22**) [52]. The results of this work showed that the copper(II) complexes showed multi-functionality including anti-inflammatory, free-radical scavenging, antibacterial and DNA/protein interaction properties. More recently, Guo and coworkers also reported similar strategy, i.e. mixed-ligand hydrazone based copper(II) complexes and examined their potential anti-lung cancer activity (**Figure 23**) [53]. Results of this study revealed that the investigated mixed-ligand copper(II) complexes especially the complex with furan pendant was found playing a good role in inducing ROS-mediated mitochondrial pathway regulated A549 cancer cell apoptosis. In both mixed ligand copper(II), the observed multi-functionality could be due to the parent complex as well as the synergistic combination of the parent hydrazone ligand as well as the co-ligands which was also observed in other works [54].

In general it have been demonstrated that hydrazone based copper (II) complexes have been extensively explored for their biological activities. Therefore, due to the vast literature report on hydrazone-based copper(II) complexes, only selected representative works are discussed in detail. From the above seminal representative works discussed, it is interesting to note that similar to the azo- based copper(II)

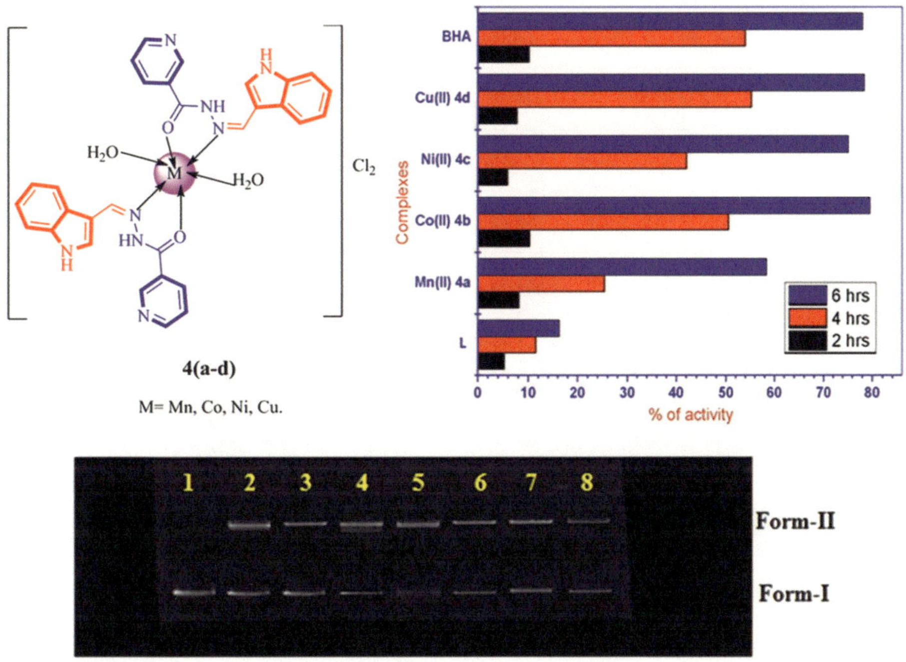

Figure 19.
Proposed structure of copper(II) complexes and percentage LDL oxidation inhibition by ligand and their metal complexes (4a–d) at 10 μM concentration (top). Cleavage of DNA (supercoiled pBR322 of 0.5 μg) by the Co(II), Ni(II) and Cu(III) complexes in a buffer at 37°C where DNA alone; DNA + 40 μl of cobalt complex + H_2O_2; DNA + 40 μl of Co(II); DNA + 60 μl of Co(II); DNA+ 40 μl of Ni(II); DNA + 60 μl of Ni(II); DNA + 40 μl of Cu(II); DNA + 60 μl of Cu(II) are lane 1–8 respectively while forms I-II are supercoiled as well as nicked circular DNA, respectively [49].

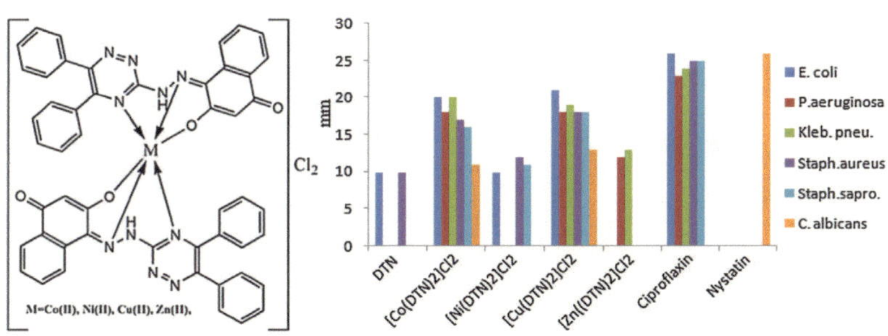

Figure 20.
Proposed structure of transition metal complexes and their antibacterial activity [50].

complexes, hydrazone-based ligands processing indol, pyrazole, nicotinic, salicylic, isatin, thiozole, quinoline and pyridoxal moieties are relatively promising for designing copper(II) complexes with pronounced biological activity (**Table 2**).

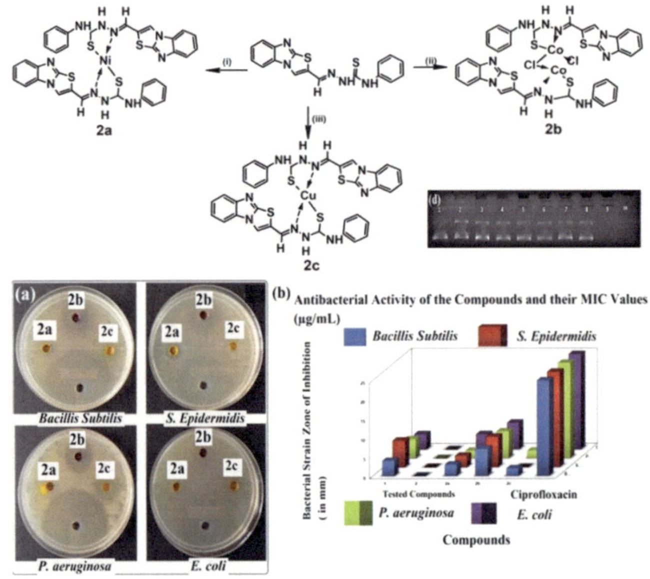

Figure 21.
Benzimidazol based hydrazone ligand and its proposed structure of transition metal complexes with inset: DNA binding where DNA control; DNA + H_2O_2; ligand + DNA; ligand + DNA + H_2O_2; Ni(II) + DNA; Ni(II) + DNA+ H_2O_2; Co(II) + DNA; Co(II) + DNA + H_2O_2; Cu(II) + DNA; Cu(II) + DNA+ H_2O_2 are lane 1-10 respectively (top). Antibacterial activity of transition metal(II) complexes with the control ciprofloxacin drug (a); and their Bar diagram representation of their antibacterial activity including the ligand (b) [51].

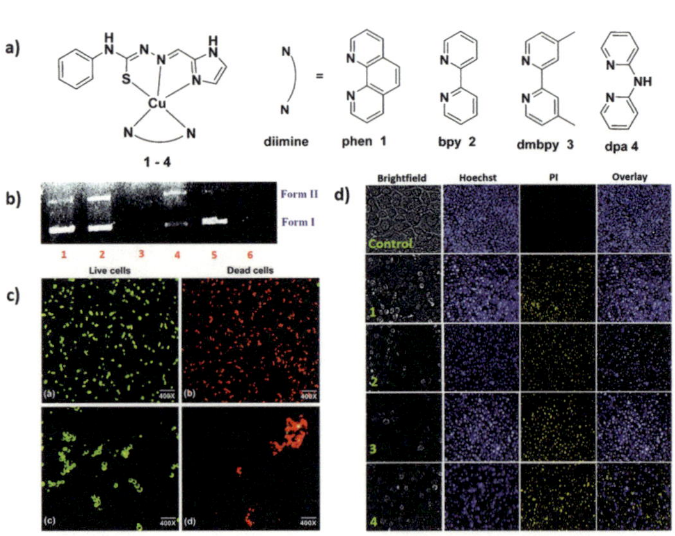

Figure 22.
Schematic representation of copper(II) complexes (1–4), the hydrazone ligand as well as the co-ligands (a). DNA cleavage (supercoiled pUC18 DNA) by copper(II) complexes 1–4 with buffer of pH = 7.14 at 37°C in the presence of ascorbic acid (ASC) as a reducing agent, where DNA control; DNA + ASC; lane 3: DNA + ASC + complex 1 (50 μM); DNA + ASC + complex 2 (200 μM); DNA + ASC + complex 3 (150 μM); DNA + ASC + complex 4 (300 μM) are lane 1–6, respectively (b). Antibacterial activity of complex 1 observed under the fluorescence microscope as well as stained using the live/dead BacLight kit against Pseudomonas aeruginosa and Staphylococcus aureus [where (a) and (c) are control and (b) and (d) are treated] (c). Hoechst 33342/PI staining of AGS human gastric cells exhibiting changes in the nuclear morphology after treating with the complexes 1–4 at concentration of 25 μM (d) [52].

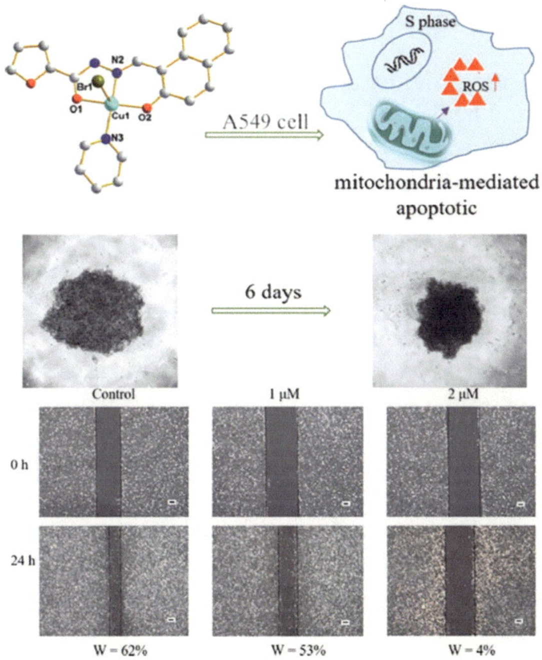

Figure 23.
Single crystal structure of the copper(II) complex and its schematic illustration of ROS-mediated mitochondrial pathway regulated A549 cancer cell apoptosis; (top) the 3D morphological changes on the A549 cells treated with the complex for 6 days with scale bar: 200 μm (middle) and wound healing results exhibited by the complex through inhibition of the A549 cells migration where pictures with scale bar: 200 μm taken at 0 h as well as 24 h, respectively (bottom) [53].

Chemical formula*	Geometry	Characterization techniques	Biological activity	Refs.
$[Cu(L_{NICO})_2]$	Octahedral	FTIR, UV–Vis, TGA, MS, XRD, SEM	Antibacterial	[55]
$[Cu(L_{INDO})(Cl)(H_2O)_2]$	Octahedral	FTIR, UV–Vis, TGA, MS, XRD, SEM, ESR	Antibacterial Antifungal and Antioxidant	[56]
$[Cu(L_{SAL})_2]$	Octahedral	FTIR, UV–Vis, MS	Antibacterial Antifungal and Antioxidant	[57]
$[Cu(L_{2QUIN})]$	Octahedral	FTIR, UV–Vis, MS, single crystal XRD	Anticancer	[58]
$[CuL_{PYRA}(phen)(CH_3OH)]$ $[CuL_{PYRA}(phen)]\cdot CH_3CH_2OH\cdot CH_3OH$	Octahedral and square pyramidal	FTIR, UV–Vis, MS, single crystal XRD	Anticancer	[59]
$[Cu(L_{ISAT})_2Cl_2]$	Octahedral	FTIR, UV–Vis, TGA, MS, EPR	Antibacterial Antifungal and DNA cleavage	[45]

**The abbreviated expressions in subscript position; NICO, SAL, INDO, QUI, PYRA, and stands for nicotinic, salicylic, indole, quinoline, pyrazole, isatin, and their derivatives respectively, for more structural details of the ligands, please refer the original article from the provided references.*

Table 2.
Copper(II) complexes of hydrazine-based ligands and their biological activities.

3. Conclusions and future outlooks

The presented chapter provides an over view of the up-to-date advances and perspectives of azo- and hydrazine-based copper(II) complexes and their biological activities. These complexes have been proved to be indispensable class of bio-active compounds with broad spectrum of biological activities such as anticancer, antibacterial, antifungal, DNA cleavage, etc. Moreover, azo- and hydrazine-based ligands bearing salicylic, thiosemicarbazide, pyridinone, pyrazole, thiazole, indole, isatin, quinoline, nicotinic, and their derivatives are found to be promising to design bio-active copper(II) compounds. Therefore, future research works are expected to explore on novel azo- and hydrazine-based copper(II) complexes owing broad spectrum biological activities accompanied with less toxicity. Biologically active starting materials such as hesperetin, curcumin and the like are recommended to rationally design the chelating azo- and hydrazine-based ligands. Additionally, chromophore tagged azo- and hydrazine-based ligands are also expected to expand the multifunctionality of their respective copper(II) bio-active complexes. Finally, the prospect of copper(II) complexes of azo- and hydrazine-based bio-active copper(II) complexes is a bright future which are expected to replace the conventional drugs which are suffering from side effects as well as pathogenic resistance. However, a collaborative work between Chemists, Biochemists as well as molecular Biologists is still needed to realize the promising biological activities of such compounds to the commercialization level. Although the research works undergoing in academia are encouraging, a collaborative work from multidisciplinary experts is a must to realize the industrialization of such auspicious compounds.

Author details

Ahlam I. Al-Sulami[1*] and Tesfay G. Ashebr[2]

1 Department of Chemistry, College of Science, University of Jeddah, Jeddah, Saudi Arabia

2 Department of Industrial Chemistry, College of Natural and Applied Sciences, Addis Ababa Science and Technology University, Addis Ababa, Ethiopia

*Address all correspondence to: 04102237@uj.edu.sa

References

[1] Santini C, Pellei M, Gandin V, Porchia M, Tisato F, Marzano C. Advances in copper complexes as anticancer agents. Chemical Reviews. 2014;**114**(1):815-862

[2] Zehra S, Tabassum S, Arjmand F. Biochemical pathways of copper complexes: Progress over the past 5 years. Drug Discovery Today. 2021;**26**(4):1086-1096

[3] Ruiz LM, Libedinsky A, Elorza AA. Role of copper on mitochondrial function and metabolism. Frontiers in Molecular Biosciences. 2021:8

[4] Bost M, Houdart S, Oberli M, Kalonji E, Huneau JF, Margaritis I. Dietary copper and human health: Current evidence and unresolved issues. Journal of Trace Elements in Medicine and Biology. 2016;**35**:107-115

[5] Denoyer D, Masaldan S, La Fontaine S, Cater MA. Targeting copper in cancer therapy: "Copper That Cancer". Metallomics. 2015;**7**(11):1459-1476

[6] Storr T, Thompson KH, Orvig C. Design of targeting ligands in medicinal inorganic chemistry. Chemical Society Reviews. 2006;**35**(6)

[7] Erxleben A. Interactions of copper complexes with nucleic acids. Coordination Chemistry Reviews. 2018;**360**:92-121

[8] Desbouis D, Troitsky IP, Belousoff MJ, Spiccia L, Graham B. Copper(II), zinc(II) and nickel(II) complexes as nuclease mimetics. Coordination Chemistry Reviews. 2012;**256**(11-12):897-937

[9] Festa RA, Thiele DJ. Copper: An essential metal in biology. Current Biology. 2011;**21**(21):R877-R883

[10] Brahma U, Kothari R, Sharma P, Bhandari V. Antimicrobial and anti-biofilm activity of hexadentated macrocyclic complex of copper(II) derived from thiosemicarbazide against Staphylococcus aureus. Scientific Reports. 2018;**8**(1):8050

[11] Gordon NA, McGuire KL, Wallentine SK, Mohl GA, Lynch JD, Harrison RG, et al. Divalent copper complexes as influenza A M2 inhibitors. Antiviral Research. 2017;**147**:100-106

[12] Malis G, Geromichalou E, Geromichalos GD, Hatzidimitriou AG, Psomas G. Copper(II) complexes with non–steroidal anti–inflammatory drugs: Structural characterization, in vitro and in silico biological profile. Journal of Inorganic Biochemistry. 2021;**224**:111563

[13] Gandra RM, Carron PM, Fernandes MF, Ramos LS, Mello TP, Aor AC, et al. Antifungal potential of copper(II), Manganese(II) and silver(I) 1,10-phenanthroline chelates against multidrug-resistant fungal species forming the Candida haemulonii complex: Impact on the planktonic and biofilm lifestyles. Frontiers in Microbiology. 2017;**8**:1257

[14] Ndagi U, Mhlongo N, Soliman ME. Metal complexes in cancer therapy – An update from drug design perspective. Drug Design, Development and Therapy. 2017;**11**:599-616

[15] Pervaiz M, Sadiq S, Sadiq A, Younas U, Ashraf A, Saeed Z, et al. Azo-Schiff base derivatives of transition metal complexes as antimicrobial agents. Coordination Chemistry Reviews. 2021;**447**:214128

[16] Gebretsadik T, Yang Q, Wu J, Tang J. Hydrazone based spin crossover

complexes: Behind the extra flexibility of the hydrazone moiety to switch the spin state. Coordination Chemistry Reviews. 2021;**431**:213666

[17] Su X, Aprahamian I. Hydrazone-based switches, metallo-assemblies and sensors. Chemical Society Reviews. 2014;**43**(6):1963-1981

[18] Kumari P, Choudhary M, Kumar A, Yadav P, Singh B, Kataria R, et al. Copper(II) Schiff base complexes: Synthetic and medicinal perspective. Inorganic Chemistry Communications. 2023;**158**(P1):111409

[19] Joshi SD, Kumar D, Dixit SR, Tigadi N, More UA, Lherbet C, et al. Synthesis, characterization and antitubercular activities of novel pyrrolyl hydrazones and their Cu-complexes. European Journal of Medicinal Chemistry. 2016;**121**:21-39. DOI: 10.1016/j.ejmech.2016.05.025

[20] Peng YX, Zhao XL, Xu D, Qian HF, Huang W. In situ metal-ion complexation and H2O2 oxidation for a pyridine-2,6-dione based disperse yellow dye. Dyes and Pigments. 2017;**136**:559-568

[21] Al-Sulami AI, Basha MT, Althagafy HS, Al-Zaydi KM, Davaasuren B, Al-Kaff NS, et al. Azo-based multifunctional molecules and their copper(II) complexes as potential inhibitors against Alzheimer's disease: XRD/Hirshfeld analysis/DFT/molecular docking/cytotoxicity. Inorganic Chemistry Communications. 2022;**142**:109535

[22] Turomsha IS, Gvozdev MY, Loginova NV, Ksendzova GA, Osipovich NP. Synthesis, characterization and biological activity of Hydrazones and their copper(II) complexes. Chemical Processing. 2023;**12**(1):73

[23] Hassan AM, Said AO, Heakal BH, Younis A, Aboulthana WM, Mady MF. Green synthesis, characterization, antimicrobial and anticancer screening of new metal complexes incorporating schiff base. ACS Omega. 2022;**7**:32418-32431

[24] Sarigul M, Sari A, Kose M, McKee V, Elmastas M, Demirtas I, et al. New bio-active azo-azomethine based Cu(II) complexes. Inorganica Chimica Acta. 2016;**444**:166-175. DOI: 10.1016/j.ica.2016.01.042

[25] Venugopal N, Krishnamurthy G, Bhojyanaik HS, Giridhar M. Novel bioactive azo-azomethine based Cu (II), Co (II) and Ni(II) complexes, structural determination and biological activity. Journal of Molecular Structure. 2019;**1191**:85-94. DOI: 10.1016/j.molstruc.2019.04.022

[26] Mallikarjuna NM, Keshavayya J, Ravi BN. Synthesis, spectroscopic characterization, antimicrobial, antitubercular and DNA cleavage studies of 2-(1H-indol-3-yldiazenyl)-4, 5, 6, 7-tetrahydro-1, 3-benzothiazole and its metal complexes. Journal of Molecular Structure. 2018;**1173**:557-566. DOI: 10.1016/j.molstruc.2018.07.007

[27] Matada MN, Jathi K, Geoffry K, Berenkere Nagarajappa R, Chander TH. Synthesis, characterization, and biological evaluation of 4-[(4-hydroxy-7-methyl-1,3-benzothiazol-2-yl) diazenyl]-5-methyl-2-phenyl-2,4-dihydro-3-pyrazol-3-one and its metal complexes. Journal of Coordination Chemistry. 2020;**73**(10):1554-1572. DOI: 10.1080/00958972.2020.1787393

[28] Maliyappa MR, Keshavayya J. Cu(II), Co(II), Ni(II), Zn(II) and Cd(II) complexes of novel azo ligand 6-hydroxy-4 methyl-2 oxo-5-[(4,5,6,7-tetrahydro-1,3-benzothiazol-2-yl)

diazenyl]-1,2-dihydropyridine 3-carbonitrile as potential biological agents: Synthesis and spectroscopic characte. Chemical Papers. 2022;**76**(6): 3485-3498. DOI: 10.1007/s11696-022-02101-7

[29] Ramashetty KB, Channabasappa PM, Seetyanaik BH, Ereshanaik RV, Nayak ANPNH, et al. Fabrication, depiction, DNA interaction, anti-bacterial, DFT and molecular docking studies of Co(II) and Cu(II) complexes of 3-methyl-1-phenyl-4-[(E)-(pyridin-2-yl)diazenyl]-1H-pyrazol-5-ol ligand. Nucleosides, Nucleotides and Nucleic Acids. 2021;**41**(1):1-22. DOI: 10.1080/15257770.2021.1991373

[30] BashaMT, Al-SulamiAI, AlthagafyHS, Alamshany ZM, Abdel-Rahman LH, Shehata MR, et al. Synthesis and characterization of novel heterocyclic diazenyl pyridinone copper(II) complexes with antimicrobial, antioxidant, and anticancer properties. Applied Organometallic Chemistry. 2023;**37**(2):1-19

[31] Lađarević J, Radovanović L, Božić B, Mašulović A, Lunić T, Radovanović Ž, et al. New copper (II) complexes derived from azo pyridone dyes: Structure characterization, thermal properties, and molecular docking studies. Applied Organometallic Chemistry. 2023;**37**(10):1-17

[32] Kasare MS, Dhavan PP, Jadhav BL, Pawar SD. Synthesis of azo Schiff Base ligands and their Ni(II), Cu(II) and Zn(II) metal complexes as highly-active antibacterial agents. ChemistrySelect. 2019;**4**(36):10792-10797

[33] Sahan F, Kose M, Hepokur C, Karakas D, Kurtoglu M. New azo-azomethine-based transition metal complexes: Synthesis, spectroscopy, solid-state structure, density functional theory calculations and anticancer studies. Applied Organometallic Chemistry. 2019;**33**(7):1-13

[34] Alaghaz ANMA, Zayed ME, Alharbi SA. Synthesis, spectral characterization, molecular modeling and antimicrobial studies of tridentate azo-dye Schiff base metal complexes. Journal of Molecular Structure. 2015;**1084**:36-45. DOI: 10.1016/j.molstruc.2014.12.013

[35] Zabin SA, Jammali M, Alzahrani AA. The bivalent Cu, Ni and Zn complexes of unsymmetrical ONO tridentate Schiff Base ligands derived from 2-aminobenzoic acid: Antimicrobial and Molluscicidal activity. Journal of Organic & Inorganic Chemistry. 2018;**4**(2):6

[36] Anitha C, Sumathi S, Tharmaraj P, Sheela CD. Synthesis, characterization, and biological activity of some transition metal complexes derived from novel Hydrazone azo Schiff Base ligand. International Journal of Inorganic Chemistry. 2011;**2011**:1-8

[37] Refat MS, El-Deen IM, Amin RR, El-Ghol S. Spectroscopic studies and biological evaluation of some transition metal complexes of a novel Schiff base ligands derived from 5-arylazo-salicyladehyde and o-amino phenol. Toxicological and Environmental Chemistry. 2010;**92**(6):1093-1110

[38] Azam M, Al-Resayes SI, Wabaidur SM, Altaf M, Chaurasia B, Alam M, et al. Synthesis, structural characterization and antimicrobial activity of Cu(II) and Fe(III) complexes incorporating azo-azomethine ligand. Molecules. 2018;**23**(4):813

[39] Refat MS, El-Deeny IM, Anwer ZM, El-Ghol S. Spectroscopic studies and biological evaluation of some transition metal complexes of Schiff-base

ligands derived from 5-arylazo-salicylaldehyde and thiosemicarbazide. Journal of Coordination Chemistry. 2009;**62**(10):1709-1718

[40] Mali SN, Thorat BR, Gupta DR, Pandey A. Mini-review of the importance of hydrazides and their derivatives-synthesis and biological activity. Engineering Proceedings. 2021;**11**(1):21

[41] Brady OL, Elsmie GV. The use of 2:4-dinitrophenylhydrazine as a reagent for aldehydes and ketones. The Analyst. 1926;**51**(599):77-78

[42] Prokai L, Szarka S, Wang X, Prokai-Tatrai K. Capture of the volatile carbonyl metabolite of flecainide on 2,4-dinitrophenylhydrazine cartridge for quantitation by stable-isotope dilution mass spectrometry coupled with chromatography. Journal of Chromatography. A. 2012;**1232**:281-287

[43] Idemudia OG, Sadimenko AP, Hosten EC. Metal complexes of new bioactive pyrazolone phenylhydrazones; crystal structure of 4-acetyl-3-methyl-1-phenyl-2-pyrazoline-5-one phenylhydrazone ampp-ph. International Journal of Molecular Sciences. 2016;**17**(5)

[44] Arora T, Devi J, Boora A, Taxak B, Rani S. Synthesis and characterization of hydrazones and their transition metal complexes: Antimicrobial, antituberculosis and antioxidant activity. Research on Chemical Intermediates. 2023;**49**(11):4819-4843. DOI: 10.1007/s11164-023-05116-1

[45] Raman N, Pothiraj K, Baskaran T. DNA-binding, oxidative DNA cleavage, and coordination mode of later 3d transition metal complexes of a Schiff base derived from isatin as antimicrobial agents. Journal of Coordination Chemistry. 2011;**64**(22):3900-3917

[46] Lodyga-Chruscinska E, Symonowicz M, Sykula A, Bujacz A, Garribba E, Rowinska-Zyrek M, et al. Chelating ability and biological activity of hesperetin Schiff base. Journal of Inorganic Biochemistry. 2015;**143**:34-47. DOI: 10.1016/j.jinorgbio.2014.11.005

[47] Sykuła A, Nowak A, Garribba E, Dzeikala A, Rowińska-Żyrek M, Czerwińska J, et al. Spectroscopic characterization and biological activity of Hesperetin Schiff bases and their Cu(II) complexes. International Journal of Molecular Sciences. 2023;**24**(1):761

[48] Gatto CC, Chagas MAS, Lima IJ, Mello Andrade F, Silva HD, Abrantes GR, et al. Copper(II) complexes with pyridoxal dithiocarbazate and thiosemicarbazone ligands: Crystal structure, spectroscopic analysis and cytotoxic activity. Transition Metal Chemistry. 2019;**44**(4):329-340. DOI: 10.1007/s11243-018-00299-8

[49] Neelufar RJ, Ankali KN, Naik N, Nuthan BR, Satish S. The Mn(II), Co(II), Ni(II) and Cu(II) complexes of (Z)-N'((1H-indol-3-yl)methylene) nicotinohydrazide Schiff base: Synthesis, characterization and biological evaluation. Journal of the Iranian Chemical Society. 2022;**19**(9):3993-4004. DOI: 10.1007/s13738-022-02580-1

[50] Priya Dharsini GR, Thanaraj C, Velladurai R. Metal chelates of tridentate (NNO) 1,2,4-Triazine Schiff Base: Synthesis, Physico-chemical investigation and pharmacological screening. Journal of Inorganic and Organometallic Polymers and Materials. 2020;**30**(7):2315-2322. DOI: 10.1007/s10904-019-01413-8

[51] Adimule V, Yallur BC, Kamat V, Krishna PM. Characterization studies of novel series of cobalt (II), nickel (II) and copper (II) complexes: DNA

binding and antibacterial activity. Journal of Pharmaceutical Investigation. 2021;**51**(3):347-359. DOI: 10.1007/s40005-021-00524-0

[52] Kathiresan S, Mugesh S, Annaraj J, Murugan M. Mixed-ligand copper(II) Schiff base complexes: The vital role of co-ligands in DNA/protein interactions and cytotoxicity. New Journal of Chemistry. 2017;**41**(3):1267-1283

[53] Wu YR, Hou LX, Lan JF, Yang F, Huang GJ, Liu W, et al. Mixed-ligand copper(II) hydrazone complexes: Synthesis, structure, and anti-lung cancer properties. Journal of Molecular Structure. 2023;**1279**:134986. DOI: 10.1016/j.molstruc.2023.134986

[54] Abebe A, Bayeh Y, Belay M, Gebretsadik T, Thomas M, Linert W. Mono and binuclear cobalt(II) mixed ligand complexes containing 1,10-phenanthroline and adenine using 1,3-diaminopropane as a spacer: Synthesis, characterization, and antibacterial activity investigations. Future Journal of Pharmaceutical Sciences. 2020;**6**(1):13

[55] Arora T, Devi J, Dubey A, Tufail A, Kumar B. Spectroscopic studies, antimicrobial activity, and computational investigations of hydrazone ligands endowed metal chelates. Applied Organometallic Chemistry. 2023;**37**(9):e7209

[56] Yernale NG, Matada BS, Vibhutimath GB, Biradar VD, Karekal MR, Udayagiri MD, et al. Indole core-based Copper(II), Cobalt(II), Nickel(II) and Zinc(II) complexes: Synthesis, spectral and biological study. Journal of Molecular Structure. 2022;**1248**:131410. DOI: 10.1016/j.molstruc.2021.131410

[57] Yernale NG, Suliphuldevara Mathada B, Shivprasad S, Hiremath S, Karunakar P, Venkatesulu A. Spectroscopic, theoretical and computational investigations of novel benzo[b] thiophene based ligand and its M(II) complexes: As high portentous antimicrobial and antioxidant agents. Spectrochimica Acta Part A: Molecular Spectroscopy. 2023;**302**(March):123114. DOI: 10.1016/j.saa.2023.123114

[58] Du Y, Chen W, Fu X, Deng H, Deng J. Synthesis and biological evaluation of heterocyclic hydrazone transition metal complexes as potential anticancer agents. RSC Advances. 2016;**6**(111):109718-109725

[59] Nurmamat M, Yan H, Wang R, Zhao H, Li Y, Wang X, et al. Novel copper(II) complex with a 4-Acylpyrazolone derivative and Coligand induce apoptosis in liver cancer cells. ACS Medicinal Chemistry Letters. 2021;**12**(3):467-476

Chapter 4

Programmed Cell Death: The Primary Bactericidal Mechanism Induced by Copper Nanoparticles

Meng-Jiun Lai, Yue-Wern Huang, Jonathan Wijaya and Betty Revon Liu

Abstract

Copper, a reddish and ubiquitous material in the world, possesses malleable and conductive properties that render copper and its alloys indispensable in vertical integration manufacturing. With advancements in nanotechnology and nanomaterials in recent decades, copper and its related nanoparticles have been engineered. Their applications include engineering, material science, photo−/electro-catalysis, biomedical drug delivery, agriculture, and antipathogen microbicides. Here, we studied the differing toxicity effects of two sizes of copper nanoparticles (CuNPs), recognized for their potent bactericidal properties. Concentration-dependent effects of both 20 and 60 nm CuNPs were significant in *Escherichia coli* (*E. coli*), *Acinetobacter baumannii* (*A. baumannii*), and *Staphylococcus aureus* (*S. aureus*). Sodium dodecyl sulfate, the dispersant of nanoparticles, caused the synergy effects with CuNPs in *A. baumannii* and *S. aureus* but not in *E. coli*. Four modulators were added to CuNP-treated bacteria. By these modulator treatments, programmed cell death was found in *E. coli*, *A. baumannii*, and *S. aureus*. By the BLAST search, caspase-related proteins were commonly identified in gut bacteria and *A. baumannii* but not in *S. aureus*. Furthermore, many proteins from *E. coli*, *A. baumannii*, and *S. aureus* were found to harbor the ULK1-catalytic domain. In short, CuNPs can be potent therapeutic agents against bacterial infections.

Keywords: membrane leakage, autophagy, apoptosis, necroptosis, cytotoxicity

1. Introduction

Copper, an earth-abundant and inexpensive material, has broad applications in manufacturing and catalysis due to its malleable and conductive properties [1]. Bulk copper is reddish and can be found in its pure form without chemical or physically controlled refining [2]. However, size-dependent color changes from brown to black show the different properties of copper at the nano-scales [3]. Bulk copper and its alloys are applied in material science, engineering, photo-electrical industries, and manufacturing. In contrast, nano-scaled copper is used in agriculture, biomedicine, environmental protection, photo-image, electronics, and superconductors [4]. Fabrication of copper nanoparticles (CuNPs) is not an emerging technology. Medieval red window glasses detected by scanning electron microscopy (SEM) and transmission electron microscopy (TEM) revealed the existence of CuNPs, which might be formed in a series process of heat and redox reactions [5]. In recent decades, various methods have been utilized for the production of copper and copper-related nanoparticles, such as copper nanoclusters and copper oxide nanoparticles. These methods include microemulsion, sol-gel techniques, green synthesis, physical smashing, chemical reduction precipitation, electro−/sono-chemical approaches, and microwave irradiation [6–8]. These copper and copper-related nanoparticles are categorized as metallic nanoparticles, containing high catalytic activities and easily interacting with organic materials like lipids and proteins [6–8]. Therefore, CuNPs are considered as potent antimicrobial agents against hazardous pathogens [9].

The increasing prevalence of new and untreatable infections has aroused concerns in recent years. Traditional antibiotics struggle with problems of invalid treatments for multidrug-resistant pathogens, and developing new strategies for these pathogen extirpations becomes an important issue [10, 11]. Gram-negative bacteria, such as *Escherichia coli* (*E. coli*) and *Acinetobacter baumannii* (*A. baumannii*), and gram-positive bacteria, *Staphylococcus aureus* (*S. aureus*), are common pathogens of opportunistic infections, which usually cause severe bacteremia and in-hospital death among hospitalized patients [12–14]. Metallic nanoparticles, containing antimicrobial properties, are reported as the workable bactericides for these three bacteria and are regarded as the last-line defense against multidrug-resistant bugs [10, 15, 16]. Various metallic nanoparticles have been studied, such as silver nanoparticles (AgNPs), CuNPs, copper oxide nanoparticles (CuONPs), and zinc oxide nanoparticles (ZnONPs), with their outstanding antimicrobial contributions and biomedical applications make them approved by the U.S. Food and Drug Administration (FDA) or undergoing clinical trials [15, 16]. Although AgNPs and CuNPs are the most commonly applied in biomedical fields, CuNPs display advantages in lower cost and higher biocompatibility [15]. There are lower risks of accumulations for CuNPs in human bodies because copper-transporting adenosine triphosphatases (Cu-ATPases) help control copper homeostasis and export excess copper out of the body [7, 17]. Our previous study also demonstrated that CuNPs were harmless to human normal skin fibroblast *in vitro* [15].

Recently, the mechanisms of bactericide in copper-based nanoparticles have been intensely discussed. Take most metallic nanoparticles, for example. Free ions released from metal, which lead to the production of reactive oxygen species (ROS), are the primary factor for killing bacteria [18]. Copper-based nanoparticles, a type of metallic nanoparticles encompassing CuO/Cu_2O, CuS, Cu^0, and other core/shell nanoparticles formed from various materials, are known to generate ROS [9, 16]. This is particularly reasonable for them as they exhibit +1 and + 2 oxidation states which enable them to interact with cell membranes through electrostatic attraction [16, 19]. In contrast, CuNPs exist in Cu^0 reduction states without charges, which should theoretically make it difficult to generate ROS. Nevertheless, Lia et al. demonstrated that the antimicrobial properties of CuNPs (Cu^0) were from ROS [15]. Additionally, Chatterjee et al. indicated that CuNPs released nascent ions via redox reactions when CuNPs were exposed to aqueous conditions, such as cell medium and serum [20]. CuNPs may employ "dry" or "wet" methods for accumulation and dissociation to reduce the proton motive force across cell membranes and increase permeability [9]. Once CuNPs entered cells through damaged cell membranes, ROS-induced oxidative stress led to lipid peroxidation, protein denaturation, and DNA degradation [9, 15, 20]. These series of attacks increased the efficacy of antimicrobial abilities in CuNPs.

Our previous study on CuNPs corroborates these findings regarding ROS-induced results [15]. We specifically investigated CuNP toxicities in *E. coli* and elucidated the correlation among CuNP sizes, concentrations, and antimicrobial activities. We found that different sizes and concentrations of CuNPs induced various mechanisms of bacterial cell death in *E. coli* [15]. *E. coli*, *A. baumannii*, and *S. aureus* are three major pathogens associated with nosocomial infections [12–14]. Studies on the antimicrobial activities of CuNPs against these three pathogens hold clinical significance. In this study, we used two sizes of CuNPs, 20 nm, and 60 nm, to investigate the bactericidal abilities and the programmed cell death mechanisms across the three pathogens.

2. Materials and methods

2.1 Nanoparticle preparation

Two sizes of CuNPs, 25 nm (marked as 20 nm) and 60–80 nm (marked as 60 nm), were purchased from Sigma-Aldrich Company (Sigma-Aldrich, St. Louis, MO, USA). As a dispersant, 1.0 mM of sodium dodecyl sulfate (SDS) (Sigma-Aldrich) was applied. The aggregation of CuNPs was prevented by capping with SDS and ultrasonic bath at 40°C for 30 min. The aqueous CuNPs were freshly prepared and used immediately. To ensure the sizes and the physicochemical properties of CuNPs, different equipment, such as transmission electron microscope (TEM; H-7500; Hitachi, Tokyo, Japan), liquid particle attractor (FlowVIEW, Hsinchu, Taiwan), Flow AOI (FlowVIEW), and Zetasizer Nano ZS (MalvernInstruments, Worcestershire, UK), were used in this study [15].

2.2 Bacterial cell culture

E. coli (Migula) Castellani and Chalmers strain 25,922 and *A. baumannii* Bouvet and Grimont strain were purchased from the American Type Culture Collection (ATCC, Manassas, VA, USA). *S. aureus*, initially derived from ATCC 13567, was purchased from the Bioresource Collection and Research Center (BCRC, Hsinchu, Taiwan). Three bacteria were cultured in Luria–Bertani (LB) broth (Becton, Dickinson and Company, Sparks, MD, USA) or LB agar (FocusBio, Miaoli, Taiwan). Bacteria were grown until the value of optical density (O.D.) at an absorbance of 600 nm reached 0.5 under aerobic conditions at 37°C with shaking at 200 rpm. The O.D. value was measured using a Bio-Rad iMark microplate reader (Bio-Rad Laboratories, California, USA).

2.3 Viability determination

To understand the antimicrobial activities of 20 and 60 nm CuNPs, three bacteria were treated with PBS (the negative control), 0 (mock), 1, 5, 10, 50, and 100 μg/mL of 20 or 60 nm CuNPs, respectively, at 37°C for 24 h. The difference between the group of negative control and mock is that mock contains dispersant while negative control does not. Bacteria were treated with 70% alcohol for 24 h as the positive control. For viability measurement, PrestoBlue® Cell Viability Reagent (Invitrogen,

Carlsbad, CA, USA) was added to each treated bacterium. After 2 h of incubation at 37°C and 200 rpm, the fluorescence at 590 nm was measured by a Varioskan LUX multimode microplate reader (ThermoFisher Scientific, MA, USA). The value of fluorescence represented the survival of cells.

To detect the minimum bactericidal concentration (MBC), three bacteria were treated with either 20 nm or 60 nm of CuNPs at the concentrations of 0, 1, 5, 10, 50, and 100 μg/mL for 24 h. After overnight treatment, each group of bacteria was transferred to the LB agar plates and incubated for another overnight. The standard colony counting method was used for colony number calculation, and MBC was determined by the colony number.

2.4 Cell death mechanism measurement

To comprehend the bactericidal mechanism of 20 and 60 nm CuNPs, we chose four modulators of programmed cell death to apply in CuNP-treated bacteria. Z-VAD-FMK (marked as Z-VAD; Sigma-Aldrich, Saint Louis, MO, USA) is an apoptosis inhibitor that binds to caspase-family proteins [21]. SBI-0206965 (marked as SBI; BioVision, Milpitas, CA, USA) and wortmannin (marked as Wort; Abcam, MA, USA) are autophagy inhibitors, which work on serine/threonine kinase ULK1 and the class III phosphatidylinositol 3-kinase (PI3K), respectively [22, 23]. Necrosulfonamide (NSA; Sigma-Aldrich, Saint Louis, MO, USA) blocking mixed lineage kinase domain-like pseudokinase (MLKL) serves as the necroptosis suppressor [24]. Before the CuNP treatments, bacteria were pretreated with 100 nM Z-VAD for 30 min, 5 μM SBI for 2 h, 100 nM Wort for 30 min, or 0.5 μM NSA for 1 h. The pretreated bacteria were then incubated with 20 nm or 60 nm of CuNPs at 0, 1, 5, 10, 50, and 100 μg/mL for 24 h, containing (or not) the same modulator at the same concentration in the pretreatment stage. The cell viabilities of bacteria rescued by modulators were detected with PrestoBlue® Cell Viability Reagent. The protocol of the PrestoBlue assay was the same as previously described.

2.5 Statistical analysis

Data are expressed as mean ± standard deviation (SD). Mean values and SDs were calculated from at least three independent experiments carried out in triplicates for each treatment group. Statistical comparisons were performed by one-way ANOVA and the Student's t-test, using statistical significance at $P < 0.05$ with stars and symbols (*, #, $, and ‡) (**Figure 1**).

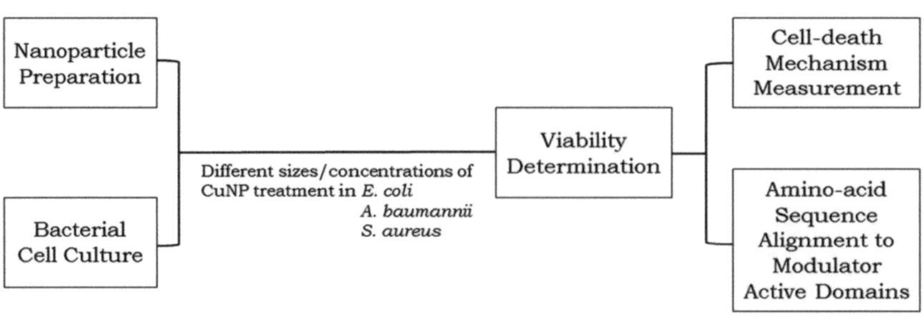

Figure 1.
The process steps in this study.

3. Results

3.1 Bactericidal activities in CuNPs

In our previous study, we disclosed the effects of different sizes and concentrations of CuNPs in *E.coli* [15]. Here, we further examined whether the same effects would be observed in other bacteria. We selected three clinically important pathogens, *E. coli, A. baumannii,* and *S. aureus*, which cover both Gram-negative and gram-positive bacteria, to investigate the bactericidal effects of 20 and 60 nm CuNPs. The bacteria were treated with different concentrations of the two sizes of CuNPs. Concentration-dependent bacterial cell viabilities and colony numbers were shown in both 20 and 60 nm CuNP treatments (**Figure 2**). Similar percentages of viabilities compared to the positive control (70% EtOH) were displayed in

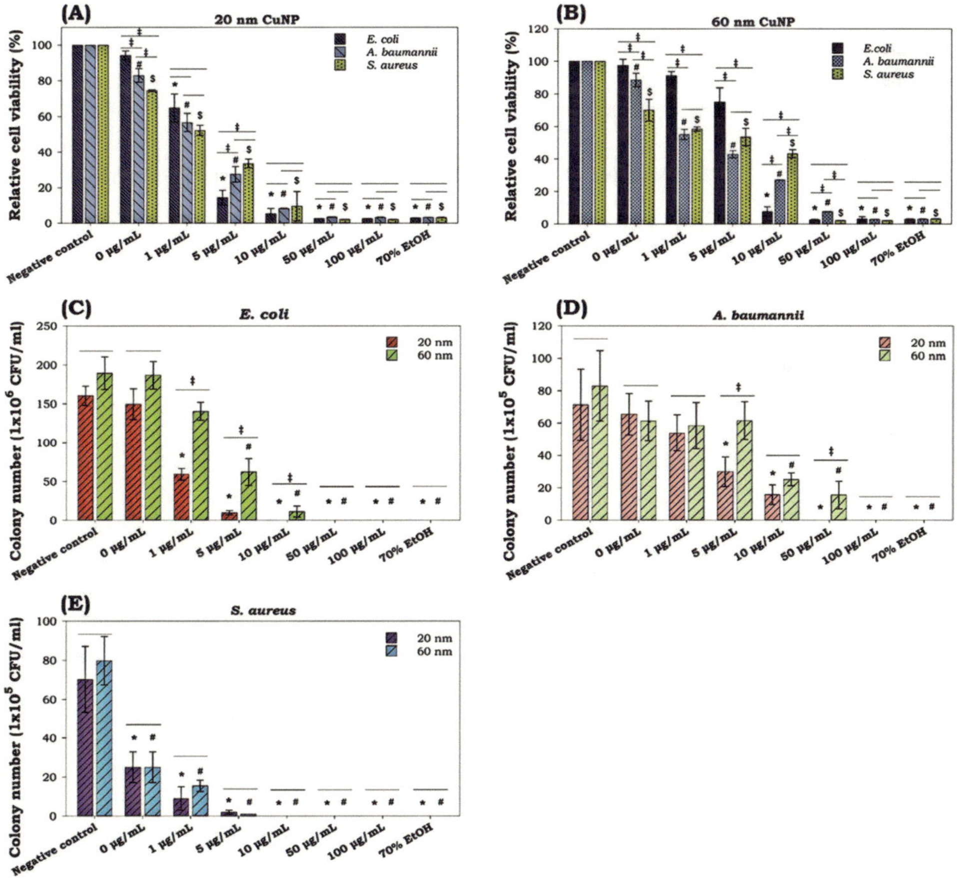

Figure 2.
*The bactericidal abilities of CuNPs in three pathogens. E. coli, A. baumannii, and S. aureus were treated with different concentrations of CuNPs in two sizes. Bacteria treated with PBS or 70% alcohol served as the negative and positive controls, respectively. The relative cell viabilities of 20 nm (A) or 60 nm CuNPs (B) in E. coli, A. baumannii, and S. aureus were determined using the PrestoBlue assay. The colony numbers were calculated for E. coli (C), A. baumannii (D), and S. aureus (E) treated with either 20 or 60 nm of CuNPs. Data are presented as mean ± SD from three independent experiments with triplicates in each treatment group. Statistical comparisons were performed by ANOVA and student t-test; *, #, and $ indicate P<0.05 for 20 or 60 nm CuNP treatments relative to the negative control, respectively, and ‡P<0.05 for comparisons among the three pathogens in the same concentration treatments.*

50 and 100 μg/mL of 20 and 60 nm CuNPs (**Figure 2A** and **B**). Significant reductions in viability were observed in the three bacteria at low concentrations (1 and 5 μg/mL) of 20 nm CuNPs. Notably, *E. coli* exhibited greater sensitivity to 5 μg/mL of 20 nm CuNPs compared to *A. baumannii* and *S. aureus* (**Figure 2A**). In contrast, mild reductions of cell viabilities were observed at 1 and 5 μg/mL of 60 nm CuNP treatments in *E. coli*. (**Figure 2B**). Significant effects of 60 nm CuNPs were evident when the concentration reached 10 μg/mL or above. Unlike *E. coli*, remarkable effects of 60 nm CuNPs in *A. baumannii* and *S. aureus* (**Figure 2B**) were observed starting from the concentration of 1 μg/mL. The most obvious difference in cell viability among the three bacteria was seen at the concentration of 10 μg/mL in 60 nm CuNP treatments.

To determine the minimum bactericidal concentration (MBC) in CuNP treatments, three bacteria were incubated with either 20 or 60 nm CuNPs, and a standard colony counting assay was performed (**Figure 2C-E**). In *E. coli* treatments, 20 nm CuNPs exhibited an excellent inhibition of colony formation at concentrations of 1, 5, 10, 50, and 100 μg/mL, while 60 nm CuNPs displayed significant inhibition at concentrations of 5, 10, 50, and 100 μg/mL (**Figure 2C**). The MBC was determined to be 10 and 50 μg/mL in 20 and 60 nm CuNP treatments, respectively (**Table 1**). In *A. baumannii* treatments (**Figure 2D**), higher concentrations of 20 and 60 nm CuNPs were required for inhibition of colony formation. The MBCs reached 50 and 100 μg/mL in 20 and 60 nm CuNP treatments, respectively (**Table 1**). *S. aureus* displayed the highest sensitivity to both sizes of CuNPs (**Figure 2E** and **Table 1**). Notably, the results in mock groups (0 μg/mL) suggested a synergy effect between the dispersant, SDS, and CuNPs in *S. aureus* treatments (**Figure 2A**, **B**, and **E**).

3.2 Bactericidal mechanism studies in CuNPs

To understand the bactericidal mechanisms triggered by CuNPs, four modulators, SBI, Z-VAD, NSA, and Wort, were chosen for subsequent experiments. These modulators are frequently employed in research on mammalian cell death mechanisms. SBI binds to Unc-51-like kinase (ULK) and interrupts the initiation of autophagy [22], while Z-VAD binds to caspase-like family proteins and inhibits apoptotic programmed cell death [21]. NSA interacts with MLKL to prevent necroptic cell death [24], and Wort increases survival rates by depleting autophagosome formation [23]. In our experiments, *E. coli*, *A. baumannii*, and *S. aureus* were pretreated with the four modulators before co-treatment with the modulators and two sizes of CuNPs, respectively (**Figure 3**).

	MBC (μg/mL)	
	20 nm CuNPs	**60 nm CuNPs**
E. coli	10	50
A. baumannii	50	100
S. aureus	10	10

Table 1.
The MBC in two sizes of CuNPs for three bacteria.

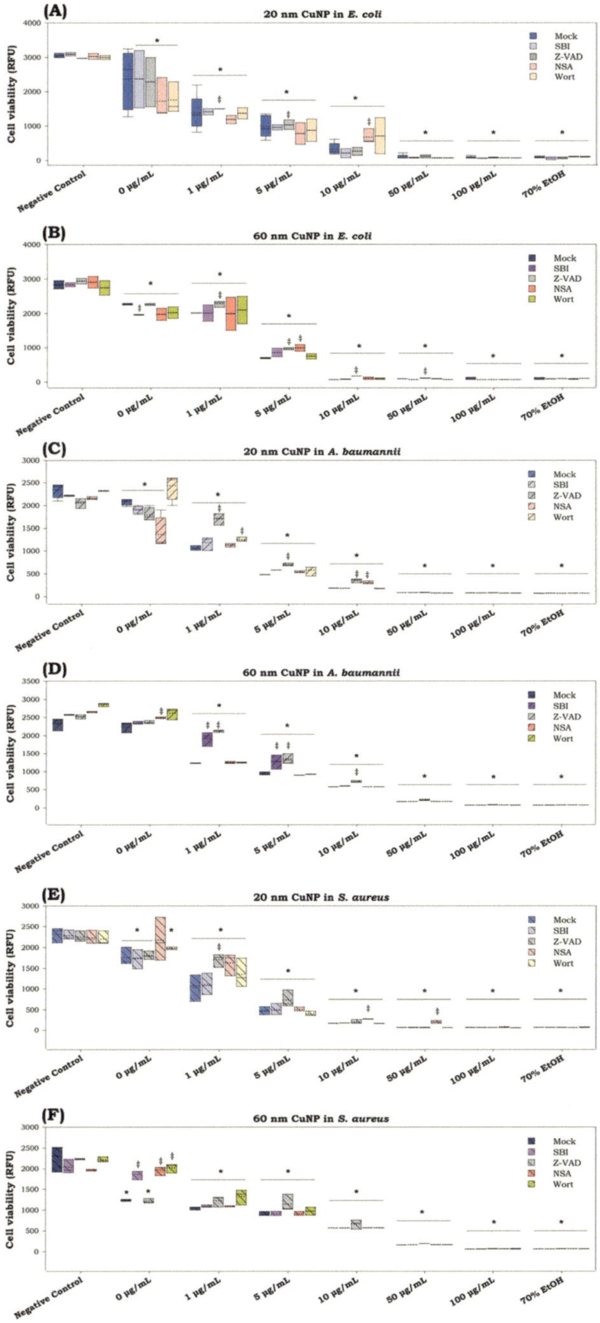

Figure 3.
Effects of CuNPs and programmed cell death modulators on bacterial cell viability. E. coli (A and B), A. baumannii (C and D), and S. aureus (E and F) were treated with PBS (negative controls) or varying concentrations (0, 1, 5, 10, 50, 100 µg/ml) of 20 or 60 nm CuNPs, either alone or in combination with programmed cell death modulators, including SBI, Z-VAD, NSA, and Wort, respectively. Cell viability was assessed by PrestoBlue assay. Statistically significant differences at P<0.05 (vs. negative control) were determined using Student's t-test. ANOVA was performed for statistical comparisons between the mock and other modulators at each concentration, with significance indicated by ‡P<0.05.*

In *E. coli*, Z-VAD- and NSA rescued the survival rate in 20 nm CuNP-treated bacteria, indicating that the cell death pathways rescued by these two modulators were the major mechanisms at low (1 and 5 μg/mL) and high concentrations (10 μg/mL), respectively (**Figure 3A**). However, the death pathway that could be rescued by Z-VAD played a vital role in all concentration treatments of 60 nm CuNPs (**Figure 3B**). Similar results to those observed in *E. coli* could also be seen in *A. baumannii* treated with 20 and 60 nm CuNPs (**Figure 3C** and **D**), with the exception of the possible involvement of the cell death pathways rescued by SBI at low concentrations (1 and 5 μg/mL) of 60 nm CuNPs (**Figure 3D**).

In contrast, *S. aureus* displayed a possible different bactericidal mechanism in CuNP treatments (**Figure 3E** and **F**). Z-VAD affected bacterial viability only at 1 μg/mL of 20 nm CuNP group, while NSA increased viability at 10 and 50 μg/mL of 20 nm CuNP (**Figure 3E**). None of the four modulators were able to raise the viabilities of *S. aureus* in all concentrations of 60 nm-CuNP treatments (**Figure 3F**).

Interestingly, the mechanisms of bactericidal effects from the dispersant of nanoparticles, SDS, were distinct in *A. baumannii* and *S. aureus*. Modulators SBI, Wort, as well as NSA rescued bacterial viabilities in *S. aureus* incubated with SDS in the mock group (0 μg/mL) of 60 nm CuNPs (**Figure 3F**). Conversely, only NSA contributed to viability elevation when *A. baumannii* was incubated with SDS in the mock group (0 μg/mL) of 60 nm CuNPs (**Figure 3D**). However, SBI decreased viabilities in the mock group when *E. coli* was treated with 60 nm CuNPs (**Figure 3B**). No significant differences were seen in the effects of SBI, Z-VAD, NSA, and Wort treatments on *E. coli*, *A. baumannii*, and *S. aureus* when bacteria were incubated with the mock of 20 nm CuNP groups (**Figure 3A, C**, and **E**).

3.3 NCBI protein database alignment of programmed cell death modulators

To verify the effects of modulators on programmed cell death in *E. coli*, *A. baumannii*, and *S. aureus*, specific domain sequences were searched on the NCBI Protein database using BLAST (**Figure 4**). Z-VAD, a pan-caspase inhibitor, acts by irreversible binding to the catalytic site of caspase proteases. In the NCBI protein database, proteins such as WP_000709408 and WP_317300176 from *E. coli* and *A. baumannii*, respectively, were annotated as caspases. These caspases contain a Peptidase_C14 domain (pfam00656, caspase domain), which has been identified in some human proteins, such as mucosa-associated lymphoid tissue lymphoma translocation protein 1 isoform b (NP_776216). However, no significant similarity was found for *S. aureus* (**Figure 4A**).

SBI is a potent inhibitor of the kinase ULK1 [25]. Through our analysis, many proteins from *E. coli*, *A. baumannii*, or *S. aureus* were identified to share similarity to the catalytic domain of human threonine-protein kinase ULK1 (NP_003556) (**Figure 4B**). Relative to the length of human ULK1 (1050 amino acids), only 260 amino acids were found to be matched among the proteins in these three bacteria.

Despite searching for the bacterial homologs of the NSA activation domain, 4HB domain of MLKL (NP_689862.1), and PI3K-like catalytic domains for Wort in the NCBI protein database using BLAST, no hits were found in *E. coli*, *A. baumannii*, or *S. aureus* (data not shown). Nevertheless, our sequencing alignment results and the bactericidal effects shown in **Figure 3** displayed consistency.

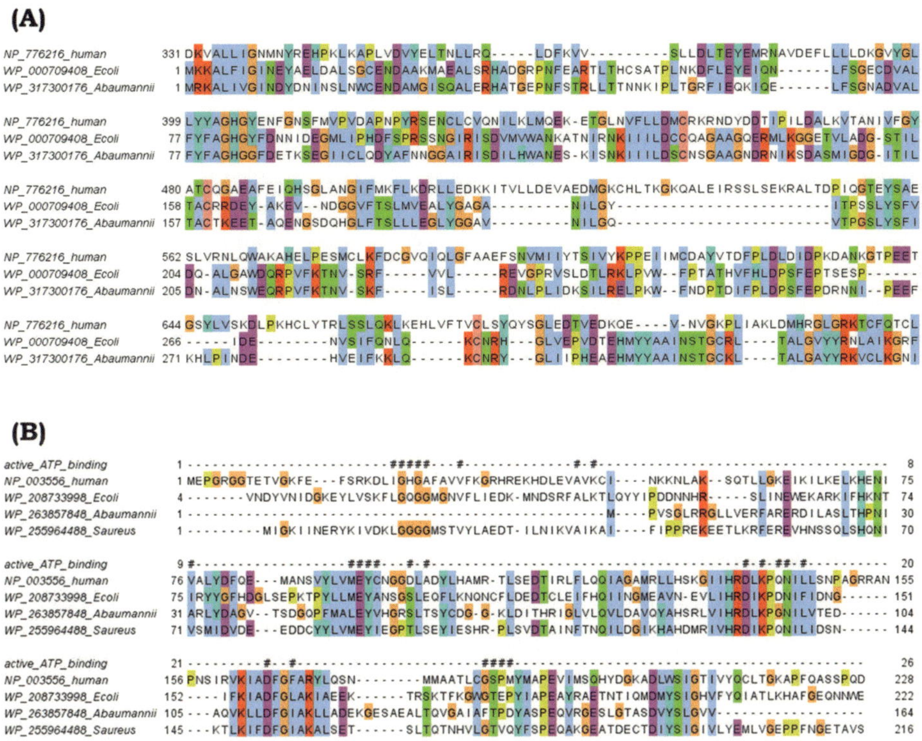

Figure 4.
Sequence alignment of the possible Z-VAD- and SBI-responsive domains in E. coli, A. baumannii, and S. aureus. (A) Alignment of the caspase domain among the caspase-like proteins from human, E. coli, and A. baumannii. (B) Alignment between the human catalytic domain of threonine-protein kinase ULK1 (NP_003556) and proteins from E. coli, A. baumannii, and S. aureus. Hash signs (#) denote possible active and/or ATP binding sites annotated in the STKc_ULK1 domain. The alignment was generated by Jalview 2.

4. Discussion

In this study, we demonstrated that both 20 and 60 nm CuNPs were effective bactericides in *E. coli*, *A. baumannii*, and *S. aureus* (**Figure 2**, **Table 1**). These nanoparticles displayed concentration-dependent antimicrobial activities. The non-toxic nature in normal human skin cells and the broad spectrum of antimicrobial activities suggest that CuNPs could be utilized in hydrocolloid dressing to defend against infections. Our investigation into the bactericidal mechanisms revealed the potential involvement of programmed cell death in these bacteria, as evidenced by the effects of different modulators and protein-sequence alignments of catalytic domains. These findings shed light on bacterial stress response and offer a promising new strategy against drug-resistant bacteria.

Metallic and metallic oxide nanoparticles have gathered interest in recent years due to their biomedical applications. They are capable of generating ROS and attacking pathogens. Our previous study confirmed this phenomenon [15], demonstrating that both 20 and 60 nm CuNPs destroyed the cell membranes of *E. coli* through ROS production. It is believed that the release of metals and electrochemical reactions from metallic-related nanoparticles, especially smaller particles with a higher surface-area-to-volume ratio, contribute to their heightened ROS production [26]. ROS can cause various cellular problems, including lipid peroxidation, protein damage, DNA breakage, blocking of carbohydrate metabolism, and inhibition of the TCA cycle [27]. As a consequence of ROS-induced damage, cellular growth, energy yields, and cell survival are reduced [28]. Some cellular organelles, such as cell membranes, mitochondria, DNA or RNA, ER systems, and some proteins or enzymes, are targets for ROS attacks [29]. Damage to cellular components can lead to cell death through necrosis, apoptosis, necroptosis, and autophagy [29]. In our previous studies, breakage of chromosomal DNA in *E. coli* treated with 20 nm CuNPs was observed by electrophoresis [15]. Furthermore, as the concentration increased, so did the damage to DNA. However, surprisingly, condensed chromosomal DNA was observed in *E. coli* treated with 60 nm CuNPs [15]. Boa and Dwyer suggested that bacteria used a specific signal transduction pathway to induce DNA condensation in the process of apoptosis when faced with ROS stress [30, 31]. Additionally, Jia et al. indicated that multiple programmed cell death pathways, such as necroptosis, ferroptosis, pyroptosis, oxeiptosis, autophagy, and apoptosis were related to ROS [32]. Therefore, it is plausible that bacterial programmed cell death is induced by ROS generated by metallic and metallic oxide nanoparticles.

Programmed cell death triggered by different stressors in *E. coli* was first discovered in 2004 [33]. In eukaryotes, apoptotic programmed cell death is characterized by DNA fragmentation, phosphatidylserine externalization, and membrane depolarization. In *E. coli*, two programmed cell death systems have been identified: the EDF-mazEF-mediated death pathway and apoptotic-like death (ALD) pathway. These pathways involve a series of signal transduction routes that control bacterial survival, ultimately leading to DNA breakage, phosphatidylserine externalization, and membrane depolarization [34]. Notably, there is a crosstalk between the EDF-mazEF-mediated death and ALD pathways. Comparing *E. coli* to eukaryotes, it emerges that multiple programmed cell death pathways found in mammalian cells, such as necroptosis, ferroptosis, pyroptosis, oxeiptosis, autophagy, and apoptosis, are highly related to ROS [35]. These programmed cell death pathways also exhibit crosstalk in their signal transduction pathways, regulated by various checkpoints [35–38]. Some signal transduction pathways involved in their survival-or-death processes share common ancestor proteins [36–38]. Different modulators or inhibitors have demonstrated to halt these progresses at specific steps [39, 40]. In this study, we used four modulators that show critical roles in necroptosis, apoptosis, and autophagy in mammalian cell studies. Co-treatment of *E. coli* and *A. baumannii* with Z-VAD and both sizes of CuNPs (**Figure 3A-D**) resulted in increased cell survival. Additionally, we observed changes in cell viabilities in *A. baumannii* co-treated with SBI and low concentrations of 60 nm CuNPs, as well as *S. aureus* co-treated with SBI and dispersant (**Figure 3D** and **F**). These results suggest that bacteria possess active domains similar to caspase-related and ULK1 proteins and respond to these modulators. Amino-acid sequence alignments support these findings (**Figure 4**). Although cell viabilities of bacteria also increased in CuNP-NSA co-treated groups, no similarity in amino-acid sequences of 4HB domain of MLKL was found in *E. coli*, *A. baumannii*, or *S. aureus*.

There are many similarities between prokaryotes and eukaryotes, yet certain properties have long been accepted as essential eukaryotic characteristics that distinguish them from prokaryotes. For instance, endocytosis was traditionally considered a primary survival behavior in eukaryotes. However, discoveries such as cyanobacteria's ability to engage in energy-dependent endocytosis challenged this notion [41]. Subsequent research revealed that diverse gram-positive bacteria like *Arthrobacter ilicis* D-50, gram-negative bacteria such as *E. coli*, and archaea like *Thermus aquaticus* possess the ability to uptake materials through macropinocytosis [42]. This process was shown to be triggered by cell-penetrating peptide-mediated protein transduction [42]. Similarly, while cytoskeletons were once believed to be exclusive to eukaryotes, proteins like FtsZ in Bacillus and MreB in *Caulobacter crescentus* exhibit similarities to actins in eukaryotes [43–45]. Additionally, the CreS protein in *C. crescentus* serves analogous functions to intermediate filaments in eukaryotes [43, 45]. Programmed cell death in bacteria, as discussed earlier, is another example of shared biological processes. In this study, we unveiled the potential for multiple programmed cell death pathways in *E. coli*, *A. baumannii*, and *S. aureus* induced by different sizes and concentrations of CuNP treatments. While our understanding is currently limited by the responses observed in modulator treatments and information available in the protein databank, the possibility of necroptosis, autophagy, or other programmed cell death pathways cannot be discounted. Understanding the regulation of cell survival-death and signal transduction pathways may hold the key to developing strategies targeting bacterial stress responses, thereby overcoming antibiotic resistance. This deeper insight into bactericidal mechanisms could pave the way for novel approaches to combat multiple drug-resistant bacteria.

5. Conclusion

Metallic and metallic oxide nanoparticles, including CuNPs, have been shown to reduce viabilities and colony formation abilities in *E. coli*, *A. baumannii*, and *S. aureus*. Nanoparticle dispersants and nanoparticles exhibit a synergic effect in killing bacteria, particularly *S. aureus*. These bacteria displayed size- and concentration-dependent sensitivities to CuNPs. Furthermore, four programmed cell death modulators elicited different responses at varying concentrations of 20 and 60 nm CuNPs. Through a search of the catalytic domains responding to the modulators in the protein database, some similarities between prokaryotes and eukaryotes were found. This study suggests the possibility of a programmed-cell death system in prokaryotes, though further research is needed to validate this concept.

Acknowledgements

We are grateful for the support from the Core Facility Center, Tzu Chi University, Taiwan. This work was supported by the Grant No. MOST 106-2813-C-320-005-B (to L.-I. T. and B. R. L.) from the National Science and Technology Council, Taiwan.

Conflict of interest

The authors declare no conflict of interest.

Appendices and nomenclature

A. baumannii	*Acinetobacter baumannii*
AgNPs	silver nanoparticles
ALD	apoptotic-like death
C. crescentus	*Caulobacter crescentus*

Cu-ATPases	copper-transporting adenosine triphosphatases
CuNPs	copper nanoparticles
CuONPs	copper oxide nanoparticles
E. coli	*Escherichia coli*
EtOH	alcohol
FDA	U.S. Food and Drug Administration
LB	Luria–Bertani
MBC	minimum bactericidal concentration
MLKL	mixed lineage kinase domain-like pseudokinase
NSA	necrosulfonamide
O.D.	optical densities
PI3K	phosphatidylinositol 3-kinase
ROS	reactive oxygen species
S. aureus	*Staphylococcus aureus*
SBI	SBI-0206965
SD	standard deviation
SDS	sodium dodecyl sulfate
SEM	scanning electron microscopy
TEM	transmission electron microscopy
ULK	Unc-51-like kinase
Wort	wortmannin
ZnONPs	zinc oxide nanoparticles
Z-VAD	Z-VAD-FMK

Author details

Meng-Jiun Lai[1], Yue-Wern Huang[2], Jonathan Wijaya[1] and Betty Revon Liu[1*]

1 Department of Laboratory Medicine and Biotechnology, College of Medicine, Tzu Chi University, Hualien, Taiwan

2 Department of Biological Sciences, Missouri University of Science and Technology, Rolla, MO, USA

*Address all correspondence to: brliu7447@gms.tcu.edu.tw

References

[1] Gawande MB, Goswami A, Felpin F-X, Asefa T, Huang X, Silva R, et al. Cu and Cu-based nanoparticles: Synthesis and applications in catalysis. Chemical Reviews. 2016;**116**(6):3722-3811

[2] Ma X, Zhou S, Xu X, Du Q. Copper-containing nanoparticles: Mechanism of antimicrobial effect and application in dentistry-a narrative review. Frontiers in Surgery. 2022;**9**:905892

[3] Das PE, Abu-Yousef IA, Majdalawieh AF, Narasimhan S, Poltronieri P. Green synthesis of encapsulated copper nanoparticles using a hydroalcoholic extract of Moringa oleifera leaves and assessment of their antioxidant and antimicrobial activities. Molecules. 2020;**25**(3):555

[4] Siddiqi KS, Husen A. Current status of plant metabolite-based fabrication of copper/copper oxide nanoparticles and their applications: A review. Biomaterials Research. 2020;**24**:11

[5] Kunicki-Goldfinger JJ, Freestone IC, McDonald I, Hobot JA, Gilderdale-Scott H, Ayers T. Technology, production and chronology of red window glass in the medieval period – Rediscovery of a lost technology. Journal of Archaeological Science. 2014;**41**:89-105

[6] Grigore ME, Biscu ER, Holban AM, Gestal MC, Grumezescu AM. Methods of synthesis, properties and biomedical applications of CuO nanoparticles. Pharmaceuticals (Basel). 2016;**9**(4):75

[7] Usman MS, El Zowalaty ME, Shameli K, Zainuddin N, Salama M, Ibrahim NA. Synthesis, characterization, and antimicrobial properties of copper nanoparticles. International Journal of Nanomedicine. 2013;**8**:4467-4479

[8] Xue Y, Cheng Z, Luo M, Hu H, Xia C. Synthesis of copper nanocluster and its application in pollutant analysis. Biosensors (Basel). 2021;**11**(11):424

[9] Ramos-Zúñiga J, Bruna N, Pérez-Donoso JM. Toxicity mechanisms of copper nanoparticles and copper surfaces on bacterial cells and viruses. International Journal of Molecular Sciences. 2023;**24**(13):10503

[10] Ermini ML, Voliani V. Antimicrobial nano-agents: The copper age. ACS Nano. 2021;**15**(4):6008-6029

[11] Mwangi J, Kamau PM, Thuku RC, Lai R. Design methods for antimicrobial peptides with improved performance. Zoological Research. 2023;**44**(6):1095-1114

[12] Laupland KB, Gregson DB, Flemons WW, Hawkins D, Ross T, Church DL. Burden of community-onset bloodstream infection: A population-based assessment. Epidemiology and Infection. 2007;**135**(6):1037-1042

[13] Liu X, Liu Y. Detection of plasmid-mediated AmpC β-lactamase in *Escherichia coli*. Biomedical Reports. 2016;**4**(6):687-690

[14] Sharifipour E, Shams S, Esmkhani M, Khodadadi J, Fotouhi-Ardakani R, Koohpaei A, et al. Evaluation of bacterial co-infections of the respiratory tract in COVID-19 patients admitted to ICU. BMC Infectious Diseases. 2020;**20**(1):646

[15] Lai MJ, Huang YW, Chen HC, Tsao LI, Chang Chien CF, Singh B, et al. Effect of size and concentration of copper nanoparticles on the antimicrobial activity in Escherichia

coli through multiple mechanisms. Nanomaterials (Basel). 2022;**12**(21):3715

[16] Mba IE, Nweze EI. Nanoparticles as therapeutic options for treating multidrug-resistant bacteria: Research progress, challenges, and prospects. World Journal of Microbiology and Biotechnology. 2021;**37**(6):108

[17] Lutsenko S, Barnes NL, Bartee MY, Dmitriev OY. Function and regulation of human copper-transporting ATPases. Physiological Reviews. 2007;**87**(3):1011-1046

[18] Mammari N, Lamouroux E, Boudier A, Duval RE. Current knowledge on the oxidative-stress-mediated antimicrobial properties of metal-based nanoparticles. Microorganisms. 2022;**10**(2):437

[19] Bogdanović U, Lazić V, Vodnik V, Budimir M, Marković Z, Dimitrijević S. Copper nanoparticles with high antimicrobial activity. Materials Letters. 2014;**128**:75-78

[20] Chatterjee AK, Chakraborty R, Basu T. Mechanism of antibacterial activity of copper nanoparticles. Nanotechnology. 2014;**25**(13):135101

[21] Gao Z, Deng G, Li Y, Huang H, Sun X, Shi H, et al. Actinidia chinensis planch prevents proliferation and migration of gastric cancer associated with apoptosis, ferroptosis activation and mesenchymal phenotype suppression. Biomedicine & Pharmacotherapy. 2020;**126**:110092

[22] Vahsen BF, Ribas VT, Sundermeyer J, Boecker A, Dambeck V, Lenz C, et al. Inhibition of the autophagic protein ULK1 attenuates axonal degeneration in vitro and in vivo, enhances translation, and modulates splicing. Cell Death and Differentiation. 2020;**27**(10):2810-2827

[23] Yu B, Yuan B, Li J, Kiyomi A, Kikuchi H, Hayashi H, et al. JNK and autophagy independently contributed to cytotoxicity of Arsenite combined with Tetrandrine via modulating cell cycle progression in human breast cancer cells. Frontiers in Pharmacology. 2020;**11**:1087

[24] Duan X, Liu X, Liu N, Huang Y, Jin Z, Zhang S, et al. Inhibition of keratinocyte necroptosis mediated by RIPK1/RIPK3/MLKL provides a protective effect against psoriatic inflammation. Cell Death & Disease. 2020;**11**(2):134

[25] Egan DF, Chun MG, Vamos M, Zou H, Rong J, Miller CJ, et al. Small molecule inhibition of the autophagy kinase ULK1 and identification of ULK1 substrates. Molecular Cell. 2015;**59**(2):285-297

[26] Balderrama-González AS, Piñón-Castillo HA, Ramírez-Valdespino CA, Landeros-Martínez LL, Orrantia-Borunda E, Esparza-Ponce HE. Antimicrobial resistance and inorganic nanoparticles. International Journal of Molecular Sciences. 2021;**22**(23):12890

[27] Hasanuzzaman M, Bhuyan M, Parvin K, Bhuiyan TF, Anee TI, Nahar K, et al. Regulation of ROS metabolism in plants under environmental stress: A review of recent experimental evidence. International Journal of Molecular Sciences. 2020;**21**(22):8695

[28] Liou GY, Storz P. Reactive oxygen species in cancer. Free Radical Research. 2010;**44**(5):479-496

[29] Sai DL, Lee J, Nguyen DL, Kim YP. Tailoring photosensitive ROS for advanced photodynamic therapy. Experimental & Molecular Medicine. 2021;**53**(4):495-504

[30] Bao H, Yu X, Xu C, Li X, Li Z, Wei D, et al. New toxicity mechanism of silver

nanoparticles: Promoting apoptosis and inhibiting proliferation. PLoS One. 2015;**10**(3):e0122535

[31] Dwyer DJ, Camacho DM, Kohanski MA, Callura JM, Collins JJ. Antibiotic-induced bacterial cell death exhibits physiological and biochemical hallmarks of apoptosis. Molecular Cell. 2012;**46**(5):561-572

[32] Jia S, Ge S, Fan X, Leong KW, Ruan J. Promoting reactive oxygen species generation: A key strategy in nanosensitizer-mediated radiotherapy. Nanomedicine (London, England). 2021;**16**(9):759-778

[33] Hazan R, Sat B, Engelberg-Kulka H. Escherichia coli mazEF-mediated cell death is triggered by various stressful conditions. Journal of Bacteriology. 2004;**186**(11):3663-3669

[34] Erental A, Sharon I, Engelberg-Kulka H. Two programmed cell death systems in Escherichia coli: An apoptotic-like death is inhibited by the mazEF-mediated death pathway. PLoS Biology. 2012;**10**(3):e1001281

[35] Yan G, Elbadawi M, Efferth T. Multiple cell death modalities and their key features (review). World Academy of Sciences Journal. 2020;**2**(2):39-48

[36] Seo J, Nam YW, Kim S, Oh D-B, Song J. Necroptosis molecular mechanisms: Recent findings regarding novel necroptosis regulators. Experimental & Molecular Medicine. 2021;**53**(6):1007-1017

[37] Mariño G, Niso-Santano M, Baehrecke EH, Kroemer G. Self-consumption: The interplay of autophagy and apoptosis. Nature Reviews. Molecular Cell Biology. 2014;**15**(2):81-94

[38] Green DR, Galluzzi L, Kroemer G. Cell biology. Metabolic control of cell death. Science. 2014;**345**(6203):1250256

[39] Zhao H, Jaffer T, Eguchi S, Wang Z, Linkermann A, Ma D. Role of necroptosis in the pathogenesis of solid organ injury. Cell Death & Disease. 2015;**6**(11):e1975-e1975

[40] Chen J-L, Wu X, Yin D, Jia X-H, Chen X, Gu Z-Y, et al. Autophagy inhibitors for cancer therapy: Small molecules and nanomedicines. Pharmacology & Therapeutics. 2023;**249**:108485

[41] Liu BR, Huang YW, Lee HJ. Mechanistic studies of intracellular delivery of proteins by cell-penetrating peptides in cyanobacteria. BMC Microbiology. 2013;**13**:57

[42] Liu BR, Huang YW, Aronstam RS, Lee HJ. Comparative mechanisms of protein transduction mediated by cell-penetrating peptides in prokaryotes. The Journal of Membrane Biology. 2015;**248**(2):355-368

[43] Gitai Z, Dye N, Shapiro L. An actin-like gene can determine cell polarity in bacteria. Proceedings of the National Academy of Sciences of the United States of America. 2004;**101**(23):8643-8648

[44] Bi EF, Lutkenhaus J. FtsZ ring structure associated with division in Escherichia coli. Nature. 1991;**354**(6349):161-164

[45] Gitai Z. The new bacterial cell biology: Moving parts and subcellular architecture. Cell. 2005;**120**(5):577-586

Section 3

Energy Harvesting

Chapter 5

Application of Copper-Based Compounds in Energy Conversion and Catalysis

Zhengwang Cheng, Shengjia Li, Mei Wang and Xinguo Ma

Abstract

Due to the crisis of energy consumpticon and environmental pollution, developing high-efficiency and low-cost catalysts is especially crucial and demanded, and the related research is increasing rapidly. Between them, copper and copper-based compounds are broadly investigated, due to their excellent properties, including ability of absorbing visible light, electronic tunability through adjusting the type and ratio of the bonded element, high catalytic efficiency and recycling property, abundant in the earth, low cost and valuable facet engineering. In this chapter, we will first introduce the crystal and electronic structure of pure copper, including the bulk and various surfaces. Then, the electronic structure of copper-based compounds will be introduced, including CuO_x, CuN_x, $CuSi_x$, and so on, whose band structure can be tuned from metal to semiconductor, topological semimetal, and even superconductor. At last, the application and mechanism in catalysis will be introduced, including plasmonic catalysis, hydrogen evolution reaction (HER), oxygen evolution reaction (OER), oxygen reduction reaction (ORR), nitrogen reduction reaction (NRR), carbon dioxide reduction reaction (CO_2RR), and single-atom catalysis (SAC). We found that Cu element can be incorporated into a broad type of materials with novel electronic structures. Furthermore, Cu-based materials play a vital role in energy conversion and catalysis.

Keywords: Cu-based compounds, electrocatalysis, photo(electro)catalysis, visible light, energy conversion

1. Introduction

In the twenty-first century, the growing global energy demand is juxtaposed against the traditional energy supply system, which is heavily reliant on fossil fuels. The consumption of these fuels produce considerable amounts of greenhouse gases and pollutants, posing significant threats to both the environment and human health [1–3]. The rising consumption of fossil energy and the resulting CO_2 emissions not only degrade air quality but also contribute to the exacerbation of global warming (**Figure 1**) [4]. To address this challenge, developing clean and sustainable energy resources is crucial. Solar and hydrogen energy are considered as promising alternative energy sources, and they can even be integrated into the prospective energy supply system [5].

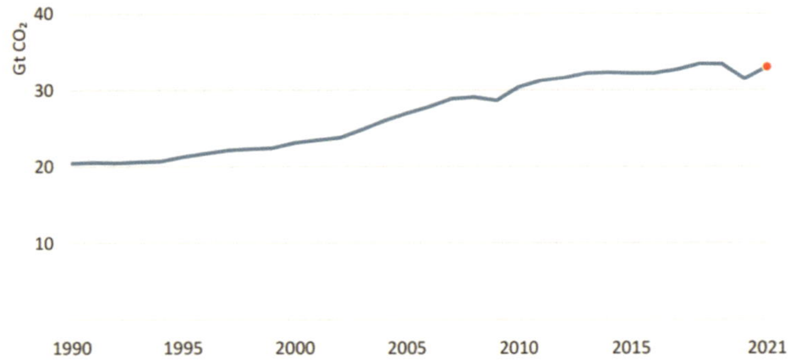

Figure 1.
Global energy-related CO_2 emissions of recent years [4].

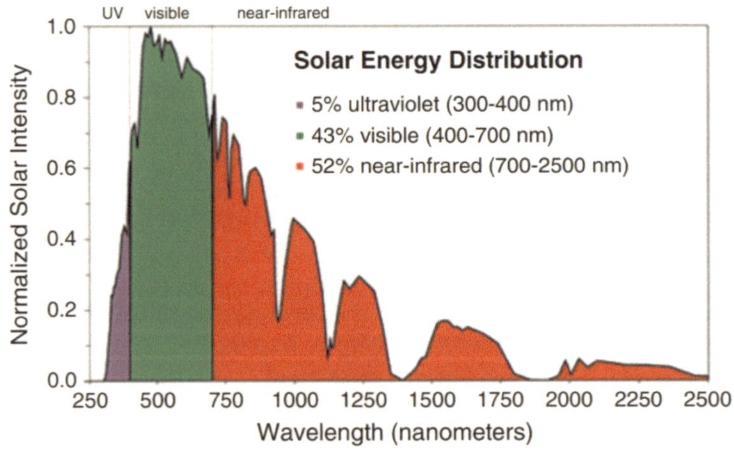

Figure 2.
Solar energy spectrogram [6].

Solar energy is a kind of clean, abundant, and widely available energy source. As depicted in **Figure 2**, the solar spectra cover a wide range from ultraviolet (UV) to near-infrared (NIR), with visible light occupying 43% [6]. Therefore, optimizing energy utilization in the visible region is imperative for enhancing solar energy utilization [7], including photovoltaic cells and solar-thermal systems, which transfer sunlight into electricity and heat. Despite achieving a certain level of solar energy utilization, these technologies face ongoing challenges, including limited conversion efficiency, high cost, and energy storage issues [8, 9].

Hydrogen (H_2), serving as a clean energy carrier, can be exclusively yielded from the consumption of water, without producing pollutants [10]. Additionally, its high-energy density renders hydrogen suitable for energy storage, transportation, and efficient utilization [11]. Hydrogen can be generated through diverse methods, encompassing water electrolysis, fossil fuel reforming, and biomass conversion. Between them, water electrolysis stands out as the cleanest method for hydrogen production, albeit being less energy-efficient and more costly [12]. Consequently, improving the energy-transfer efficiency and reducing the cost of water splitting constitute a pivotal role in maximizing the utility of hydrogen energy [13]. In addition, how to reduce the harmful gas CO_2 and NO_x is important for environmental

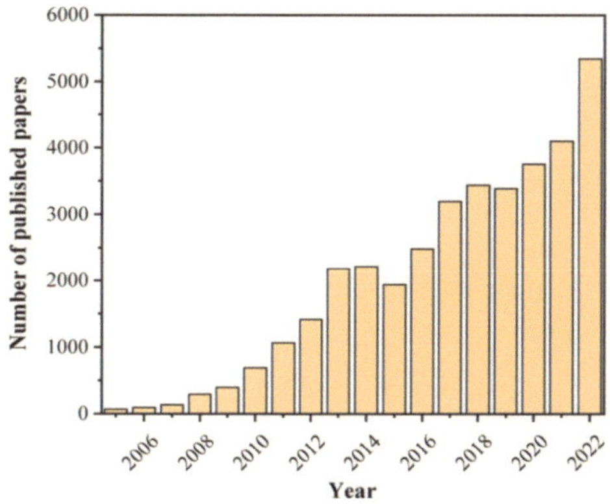

Figure 3.
Trends of published papers on solar energy utilization over the last two decades.

protection. Then, convert them into useful nitric and carbonous chemicals, such as NH_3 and C_2H_4, utilizing solar energy through photocatalysis or photoelectrocatalysis could be a preferable route.

In recent years, significant progress has been achieved in the research on the utilization of solar energy. As illustrated in **Figure 3**, the publication number of the related research has increased continually over the past two decades. However, there is still an ongoing need to improve solar energy conversion efficiency and catalysis, specifically in the visible-light range.

This chapter will first introduce the crystal and electronic structure of pure copper in Section 2, covering the bulk and low-index surfaces. In Section 3, the crystal and electronic structures of various copper-based compounds will be briefly summarized, encompassing semiconductors, alloys, topological semimetals, and superconductors. In Section 4, the main applications and mechanisms of Cu-based catalysis and energy conversion will be introduced. The overarching goal is to facilitate readers' comprehension of the crystal structure, electronic structure, applications and mechanisms of copper and copper-based compounds in energy conversion and catalysis.

2. Crystal and electronic structure of pure copper

2.1 Bulk copper

As shown in **Figure 4**, copper (Cu) possesses a face-centered cubic (FCC) crystal structure, where copper atoms reside at corners and face centers of the crystal unit cell, with the number of nearest neighboring atoms reaching the maximum 12, *that is,* closely packed. The lattice constant of bulk Cu is approximately 3.61 Å. The electrical and thermal conductivities of Cu at room temperature are 5.96×10^7 S/m and 398 W/(m·K), respectively. The electronic configuration of Cu is [Ar] $3d^{10}4s^1$, indicating a fully occupied 3d shell and semi-filled 4 s orbital, resulting in the typical metal bands, without a forbidden band gap between the valence band (VB) and conduction band (CB). Besides, Cu demonstrates specific light absorption properties. For bulk Cu, it owns good absorption in the visible spectral range and results in the typical metallic color. Besides, plasmonic resonance is a ubiquitous phenomenon in Cu, which is the

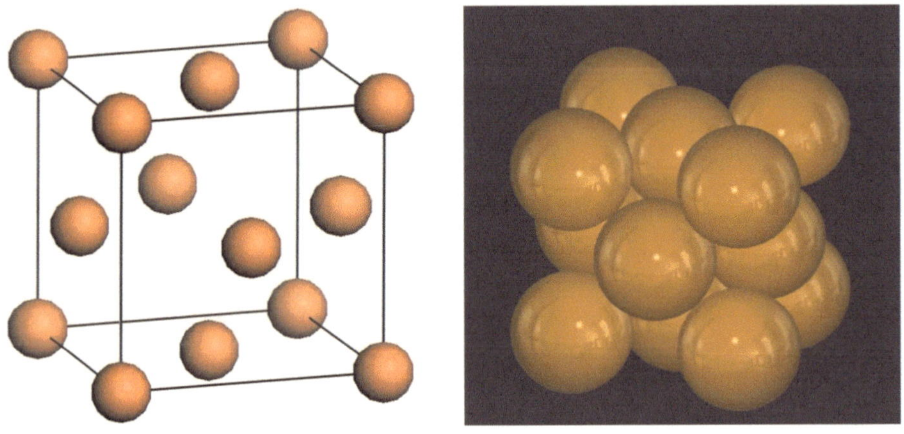

Figure 4.
Crystal structure of bulk copper.

excited states of collective oscillations of free electrons in a metal, under the irradiation of light. For Cu nanomaterials, the resonance absorption can be tuned from the ultraviolet to the visible-light region.

2.2 Low-index surface of copper

Copper can expose a variety of crystalline facets, among which the low-index facets are particularly active and have garnered significant attention. Following we will elucidate the atomic and electronic structures of several low-index Cu facets.

2.2.1 Cu(100) surface

Figure 5 illustrates the atomic and electronic structures of the Cu(100) surface [14, 15]. **Figure 5(a)** depicts the crystal structure of Cu, with the Cu(100) surface highlighted in gray on the upper panel, the lower panel shows the top-view image, and unit cell is marked with a square. **Figure 5(b)** illustrates the three-dimensional (3D) Brillouin zone (BZ) of bulk Cu and its projection onto the (100) plane. **Figure 5(c)** displays the typical scanning tunneling microscope (STM) image of Cu(100) surface, with each bright dot corresponding to a Cu atom, it illustrates an obvious relationship between the atomic arrangement in the STM image in **Figure 5(c)** and the crystal structure from **Figure 5(a)**. **Figure 5(d)** presents the theoretically calculated bands and experimentally measured ones by angle-resolved photoelectron spectroscopy (ARPES) with $h\nu = 27$ eV on the Cu(100) surface along Γ-M direction.

2.2.2 Cu(110) surface

Figure 6 illustrates the atomic and electronic structure of the Cu(110) surface [14, 16]. **Figure 6(a)** depicts the crystal structure of Cu, highlighting the Cu(110) surface in gray at the top, and displaying the corresponding atomic arrangement at the bottom, with the rectangle-shaped unit cell indicated. The corresponding BZ is

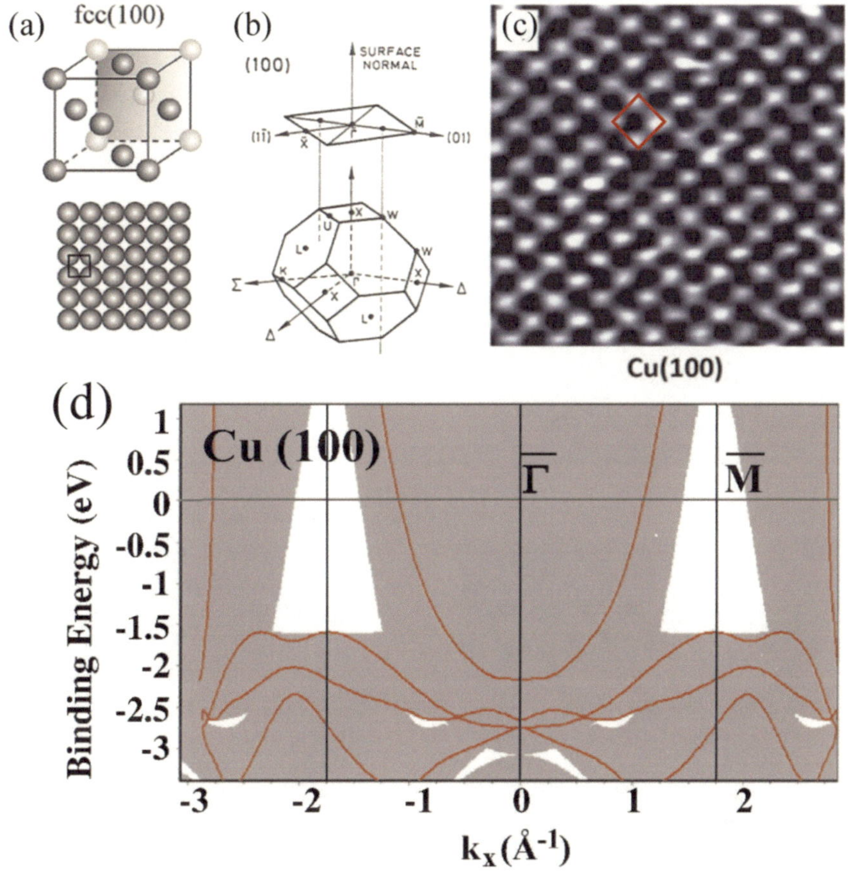

Figure 5.
(a) Crystal structure of Cu(100). (b) 3D BZ of bulk Cu and the surface projected BZ of Cu(100). (c) Typical STM image of Cu (100) [14]. (d) Tight-bond calculated bulk band for Cu(100) surface (gray shaded), and the red line represents the experimentally obtained bands by ARPES spectra with $h\nu$ = 27 eV [15].

depicted in **Figure 6(b)**. Furthermore, the STM image in **Figure 6(c)** shows a regular array of bright dots, and the period is consistent with the atomic arrangement in **Figure 6(a)**. Finally, the high-resolution surface state (SS) of Cu(110) is shown in **Figure 6(d)**, informing a parabolic dispersion.

2.2.3 Cu(111) surface

Figure 7 illustrates the atomic and electronic structures of the Cu(111) surface [14, 17]. **Figure 7(a)** depicts the crystal structure of Cu, with the Cu(111) surface highlighted in gray in the upper panel, and the top-view shown in the lower panel. **Figure 7(b)** presents the corresponding BZ. Additionally, **Figure 7(c)** displays the typical STM image of the Cu(111) face, with the atomic arrangement aligned with that in **Figure 7(a)**. Finally, **Figure 7(d)** presents the ARPES result of Cu(111) face, with the Shockley-type SS and *sp* band clearly observed.

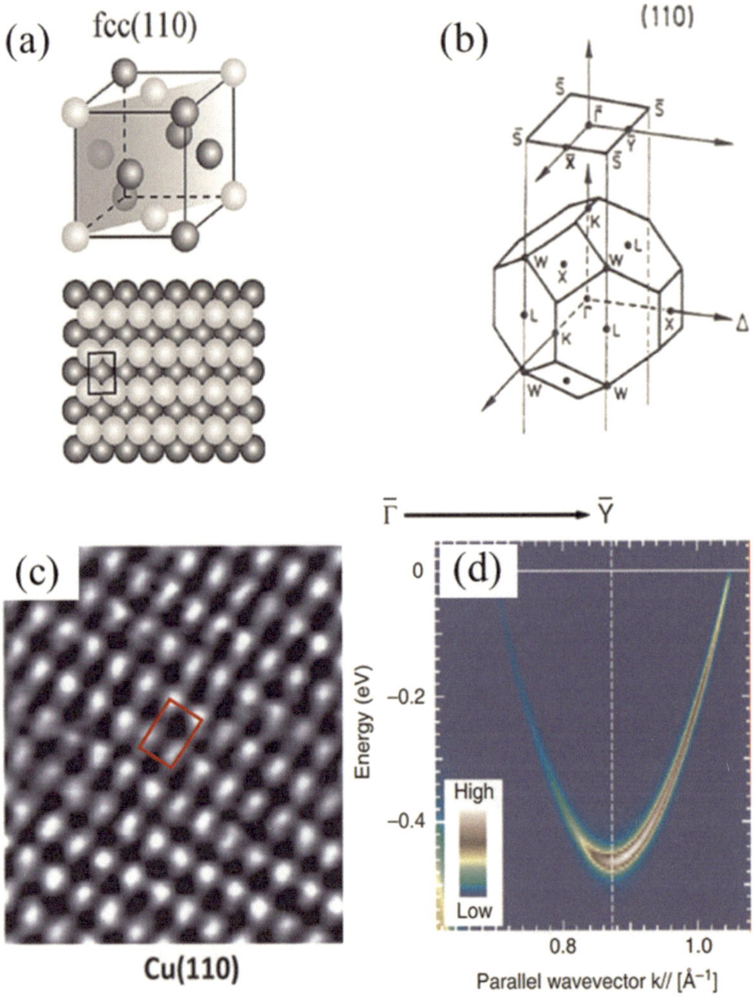

Figure 6.
(a) Crystal structure, (b) BZ, and (c) typical STM image of Cu (110) surface [14]. (d) ARPES intensity plot of Cu(110) surface state [16].

2.3 Facet-dependent catalysis of copper nanomaterials

The distinct crystalline surfaces of Cu normally own different physical and chemical properties, including the catalysis activity and the corresponding reaction pathways. Therefore, surface modulation is an important research hotspot in catalysis.

Figure 8(a) depicts the decomposition kinetics of formate (HCO_2) on Cu(111), Cu (110), and Cu(100) surfaces [18] through density functional theory (DFT, solid line) and potential energy surface (PES, dashed line), which shows great coincidence within 10 meV. It can be seen that the activation barrier heights for HCO_2 decomposition on Cu(111) are close to those of Cu(100) and lower than those of Cu(110), while the values are 1.104, 1.155, and 1.483 eV, respectively. For the mean translational energy $\langle E_{trans} \rangle$ of the CO_2 product (**Figure 8(b)**) and mean kinetic energy of adsorbed H atom (**Figure 8(c)**) during HCO_2 decomposition, the former one obtains more

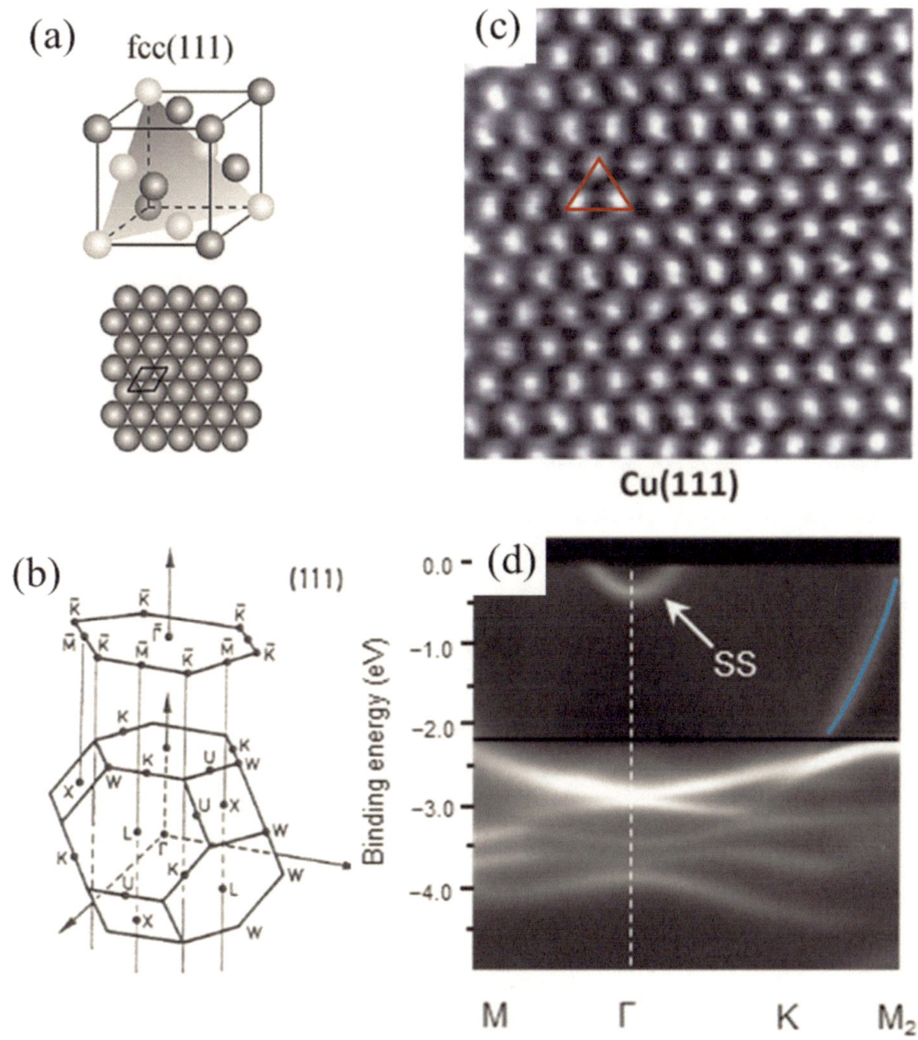

Figure 7.
(a) Crystal structure, (b) BZ, and (c) typical STM image of Cu (111) surface [14]. (d) ARPES spectra of pristine Cu(111), the SS state and sp *band are indicated with a white arrow and blue line, respectively [17].*

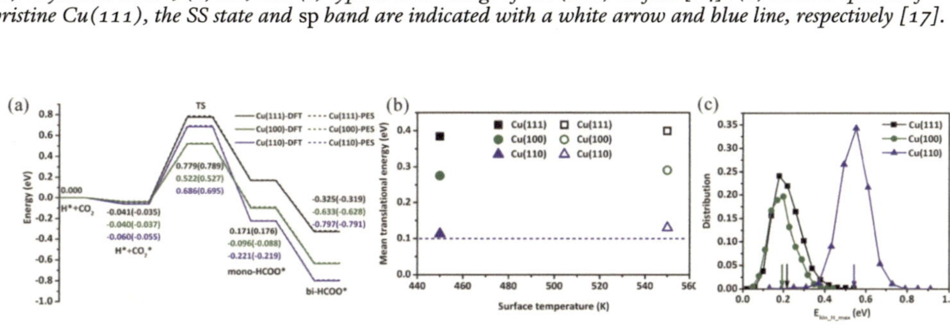

Figure 8.
(a) Comparison of stationary point energies of HCO_2 decomposition on relaxed Cu(111), Cu(100), and Cu(110) surfaces. (b) Mean translational energy E_{trans} of CO_2 product from HCO_2 decomposition. (c) Distribution of the maximum kinetic energy of adsorbed H atom ($E_{kin\_H\_max}$) during HCO_2 decomposition, the arrows indicate the mean $E_{kin\_H\_ma}$ [18].

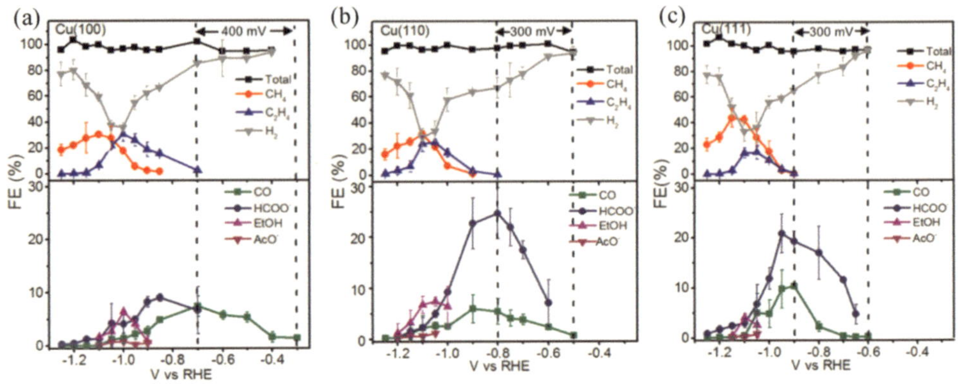

Figure 9.
Faradaic efficiencies (FE) of CO_2 reduction products formed on (a) Cu(100), (b) Cu(110), and (c) Cu(111) surfaces. The dashed lines mark the onset potentials of CO and C_2H_4 formation [19].

kinetic energy on Cu(111) and Cu(100) surfaces, but the latter one obtains most energy on Cu(110) surface. These results undoubtedly inform the facet-dependent catalysis of Cu material.

Figure 9 illustrates the electrochemical reduction of CO_2 on different facets of single crystal Cu [19]. Although the onset potentials for C_2H_4 formation are consistently 300–400 mV more negative than those for CO product on Cu(100), Cu(110), and Cu(111) surfaces, the values on each surface are different, that is, Cu(100) (−0.7 V, −0.3 V) < Cu(110) (−0.8 V, −0.5 V) < Cu(111) (−0.9 V, −0.6 V), informing a facet-dependent activity for CO_2 reduction.

3. Crystal and electronic structure of typical copper-based compounds

As we know, there are numerous copper-based compounds reported, with distinct structures and properties. This section will introduce the crystal and electronic structures of various classes of copper-based compounds, encompassing semiconductors, alloys, semimetals, and superconductors.

3.1 Copper-based semiconductor

Semiconductor is a kind of material whose electrical conductivity falls between that of a conductor and insulator. The properties of semiconductors are primarily determined by their energy band structure, separated by an <5 eV energy band gap between the valence band (VB) and conduction band (CB). At absolute zero temperature and without excitation of the external field, VB is completely occupied, and CB is entirely unoccupied, thereby preventing the flow of electrons from forming a current. However, as the temperature increases, or under the irradiation of light with enough energy, partial electrons can be excited from VB to CB, allowing the flow of current. There are numerous copper-based semiconductors, typically including CuO, Cu_2O, Cu_3N, and so on.

CuO is a kind of semiconductor with the monoclinic unit cell and space group *C2/c*, as depicted in **Figure 10(a)**. Each copper atom is surrounded by four oxygen atoms, and each oxygen atom is similarly surrounded by four copper atoms. **Figure 10(b)** illustrates the calculated band structure of CuO. The gray portion represents the band

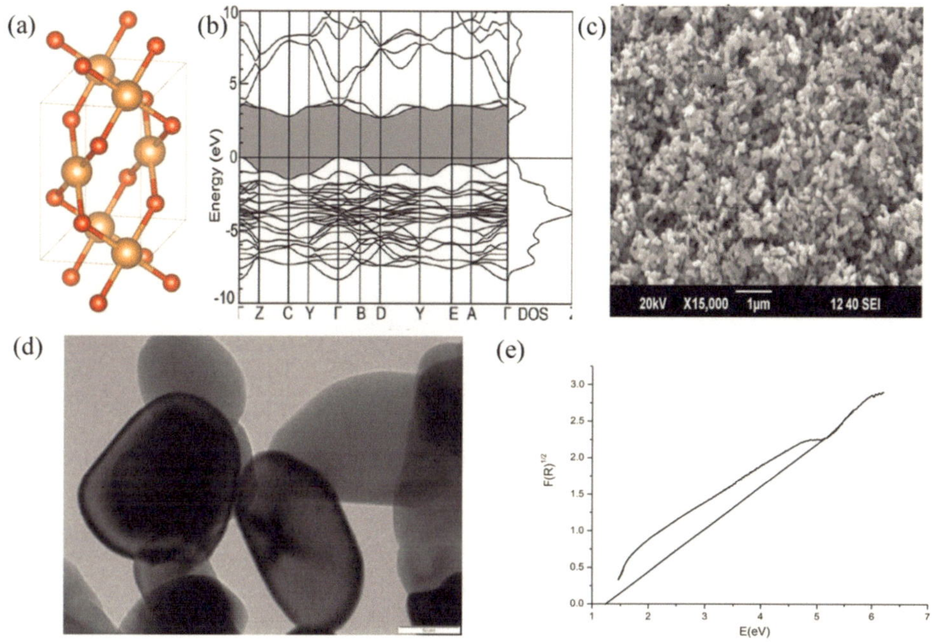

Figure 10.
(a) Crystals structure and (b) calculated electronic band structure of CuO [20]. (c) SEM image, (d) TEM image and (e) Tauc plot of CuO nanoparticles [22].

gap of Cu, and the Fermi level (E_F) located near VB, indicating a p-type semiconductor [20]. Hall effect tests reveal that CuO has a resistivity of 0.26 Ω cm, mobility of 0.12 $cm^2/V{\cdot}s$, and a carrier concentration of 7.4×10^{19} cm^{-3} [21]. **Figure 10(c–d)** displays the SEM and TEM images of CuO nanoparticles fabricated using a simple sol-gel method. The CuO nanoparticles are uniformly distributed, spherical in shape, with particle sizes ranging from 20 to 50 nm. The optical band gap of CuO varies between 1.2 and 2.1 eV, attributed to differences in the preparation process and sample quality. The band gap of the above CuO nanoparticles is estimated to be 1.2 eV (**Figure 10(e)**) [22].

The crystal structure of Cu_2O is depicted in **Figure 11(a)**. The space group of Cu_2O is *Pn3m*, and it possesses a cubic structure, with oxygen atoms located at eight vertex positions and one body center position in the cell, while four copper atoms are situated at the midpoint of the line connecting the vertices and the body center. **Figure 11(b)** illustrates the calculated band structure of Cu_2O, where the gray portion represents the band gap, and E_F is positioned near VB [20]. Hall effect tests suggest that Cu_2O is a kind of p-type semiconductor, with a resistivity of 149 Ω·cm, mobility of 51 $cm^2/V{\cdot}s$, and a carrier concentration of 1.5×10^{15} cm^{-3} [21]. The band gap of the Cu_2O film prepared by electrodeposition is 2.1 eV (**Figure 11(c)**) [23]. **Figure 11(d–f)** shows the Cu_2O films prepared *via* the magnetron sputtering (MS) method at different temperatures. As the temperature increases from 300 K to 600 K and 1070 K, the films transition from fibrous grains to columnar grains, and the grain size increases. The statistical grain sizes are 79 ± 17, 228 ± 57, and 884 ± 373 nm, respectively [24].

As shown in **Figure 12(a)**, Cu_3N possesses an anti-ReO_3 cubic structure with a space group of $\boldsymbol{Pm\bar{3}m}$ and a lattice constant of a = 3.82 Å. The Cu_3N crystal lattice is

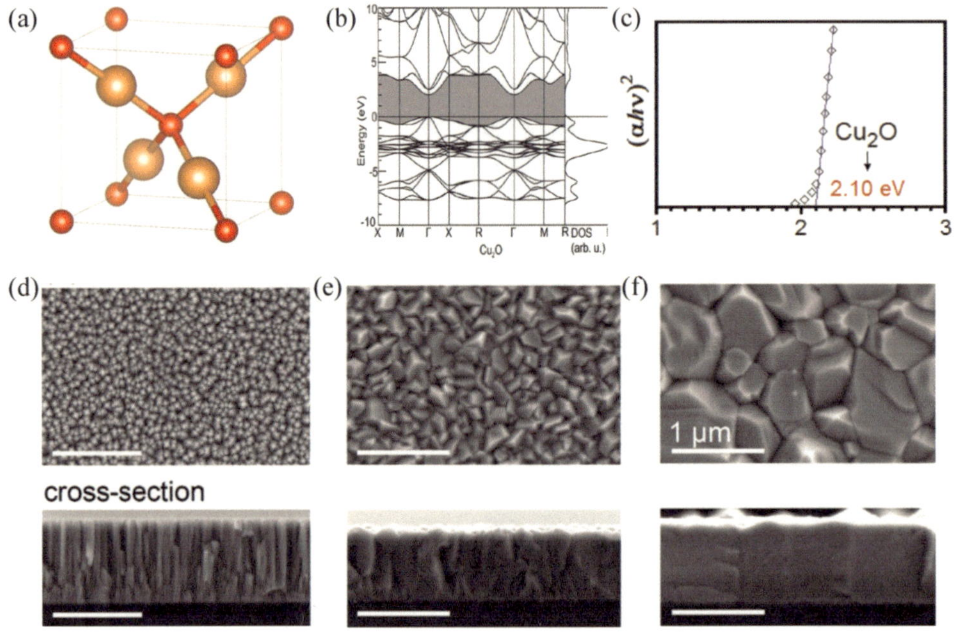

Figure 11.
(a) Crystal structure, (b) electronic band structure of Cu_2O [20]. (c) Plot of $(\alpha h\nu)^2$ vs. $h\nu$ for Cu_2O [23]. (d-f) SEM images of Cu_2O films grown at different temperatures, (d) 300 K, (e) 600 K, and (f) 1070 K [24].

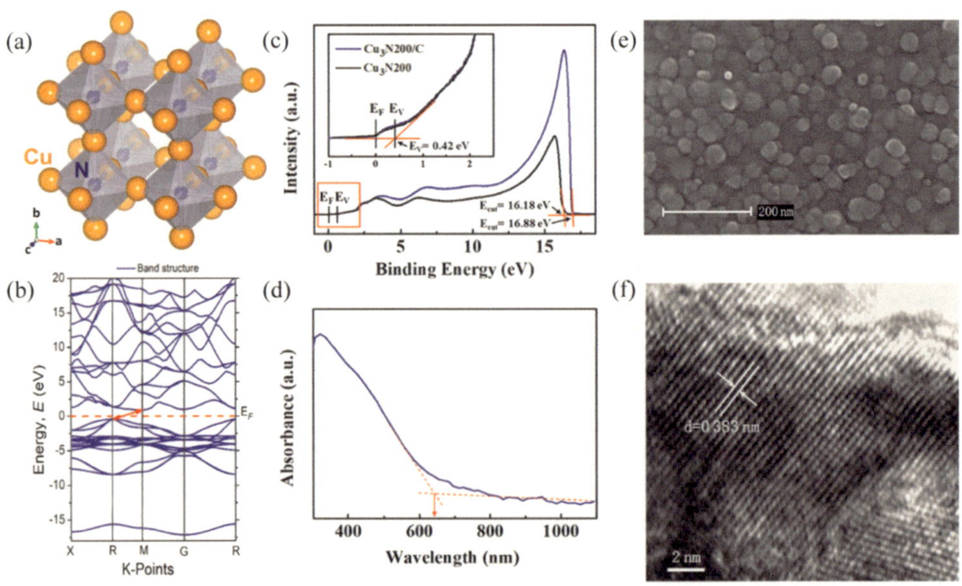

Figure 12.
(a) Crystal structure and (b) calculated band structure of Cu_3N [25, 26]. (c) UPS spectra, the inset is enlarged around E_F *to estimate the position of the valence band maximum. (d) UV–vis absorption spectrum [27]. (e) SEM and (f) high-resolution TEM images of Cu_3N [28].*

composed of Cu_6N octahedrons through vertex sharing [25]. While each N atom is surrounded by six Cu atoms, each Cu atom is shared by two N atoms. **Figure 12(b)** displays the calculated band structure of Cu_3N. The valence band maximum

(VBM) and conduction band minimum (CBM) are located at R and M points, respectively, informing an indirect band gap of 1.4 eV [26]. Ultraviolet photoelectron spectroscopy (UPS, **Figure 12(c)**) reveals that the VBM of Cu_3N located at 0.42 eV below E_F, UV-visible (UV-vis) absorption spectrum measurement (**Figure 12 (d)**) gives the optical band gap to be 1.92 eV, then CBM should be located at 1.50 eV above E_F. Comparatively speaking, pristine Cu_3N's VBM is closer to E_F than CBM, suggesting Cu_3N is a p-type semiconductor [27]. **Figure 12(e–f)** shows the SEM and high-resolution TEM images of Cu_3N prepared by MS method, revealing uniform spherical nanoparticles with an approximate size of 40 nm, while the TEM image shows (001) plane with the crystal space of 0.383 nm [28].

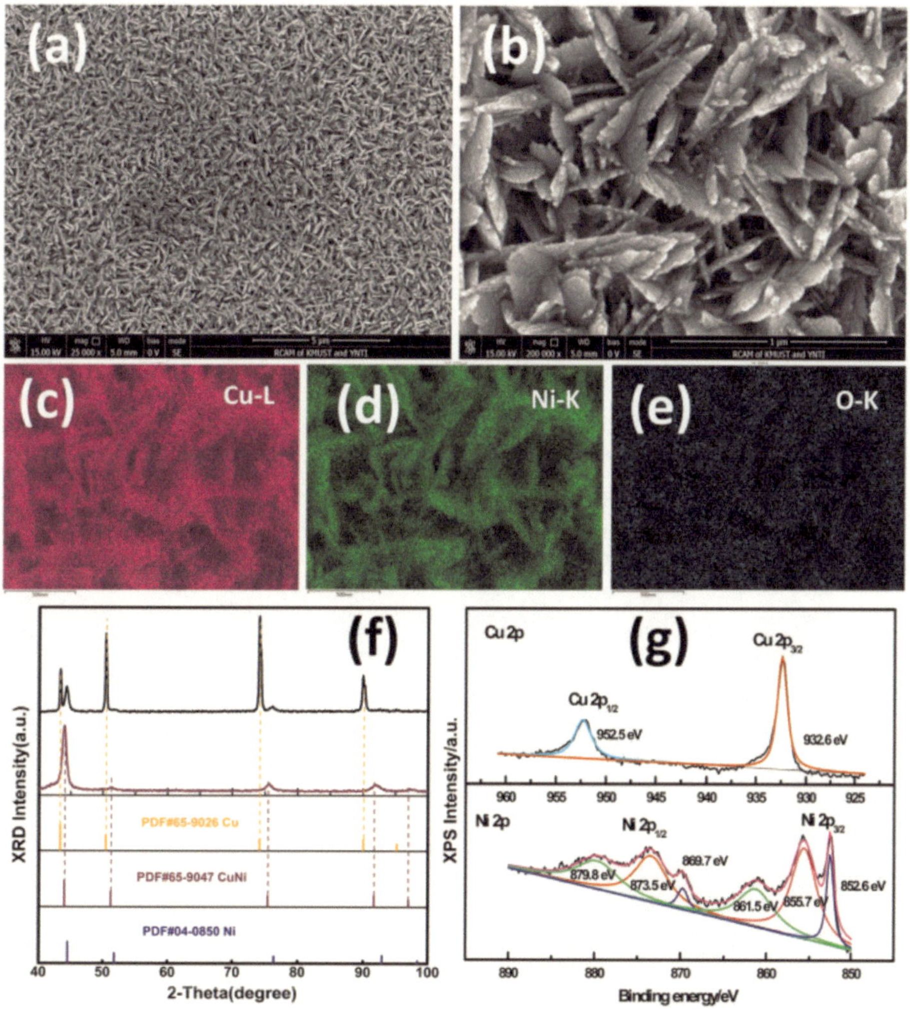

Figure 13.
(a) Low- and (b) high-magnified SEM images of $Ni_{0.7}Cu_{0.3}$ alloy film via electrodeposition. (c-e) EDX mapping of Cu, Ni, and O elements for the as-prepared sample. (f) XRD patterns and (g) XPS results of Cu 2p and Ni 2p [29].

3.2 Copper-based alloy

Copper-based alloys are kinds of composites composed of Cu and other elements, such as Zn, Al, N, Mn, Pb, and so on. They are extensively utilized due to their favorable electrical and thermal conductivity, corrosion resistance, and workability. This subsection introduces several copper-based alloys.

$Ni_{0.7}Cu_{0.3}$ alloy is a promising catalyst toward hydrogen evolution [29]. As shown in **Figure 13**, $Ni_{0.7}Cu_{0.3}$ alloy films can be prepared through a facile potentiostatic electrodeposition method (**Figure 13(f)**). They reveal a highly porous structure consisting of well-organized 3D nanosheets (SEM, **Figure 13(a–b)**), with Cu and Ni uniformly distributed (EDS mapping, **Figure 13(c–e)**), and mainly show metallic states (XPS, **Figure 13(g)**).

As depicted in **Figure 14**, $Ag_{0.8}Cu_{0.2}$ alloy, which can be applied in zinc-air batteries, was synthesized on nickel foam *via* a two-step electrochemical substitution reaction. The morphology of the $Ag_{0.8}Cu_{0.2}$ alloy exhibits a dense and uniform dendritic shape, with Ag and Cu elements uniformly distributed. The clear lattice stripes in HRTEM image and sharp diffraction spots suggest a high crystallinity [30].

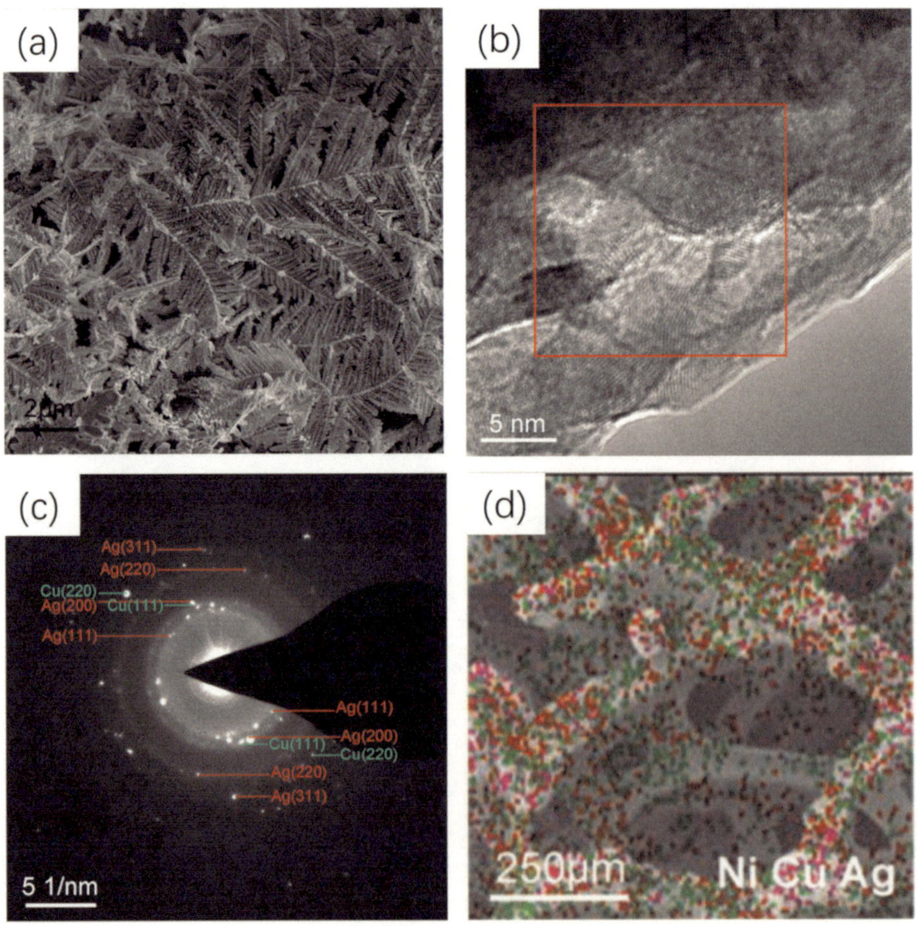

Figure 14.
(a) The FE-SEM image, (b) HRTEM image, (c) SAED pattern, and (d) EDS overlay image of the $Ag_{0.8}Cu_{0.2}$ *catalyst [30]. Red, purple, and green colors indicate Cu, Ag, and Ni, respectively.*

Except for the binary Cu-based alloy, ternary Cu-based alloy is another broadly applied kind. For example, **Figure 15** illustrates the structure of ternary CoFeCu alloy on Ni foam (NF), an efficient catalyst for oxygen evolution reaction (OER, for more information refer to Section 4.3). As the copper content increases from 0 mmol to 40 mmol, the morphology changes from nanosheets to nanoparticles, branched structures, and dendritic morphology (**Figure 15(a–d)**) [31]. The TEM and elemental mappings suggest a uniform distribution of Co, Fe, and Cu elements. Besides, the crystal structure of the CoFeCu alloy is schematically depicted in **Figure 15(e)**.

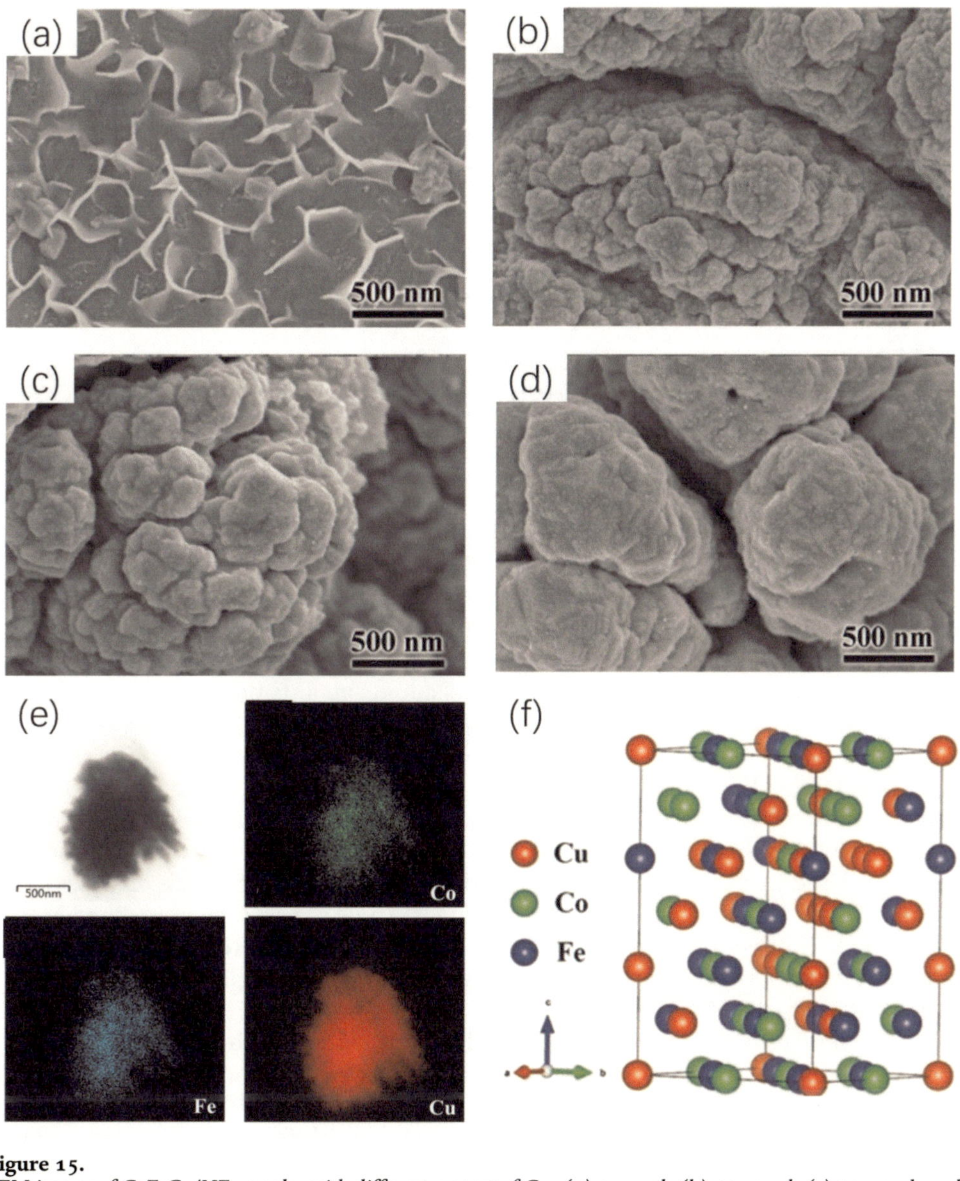

Figure 15.
SEM images of CoFeCu/NF samples with different content of Cu, (a) 0 mmol, (b) 10 mmol, (c) 20 mmol, and (d) 40 mmol. (e) TEM image and corresponding elemental distribution images and (f) schematic crystal structure of CoFeCu [31].

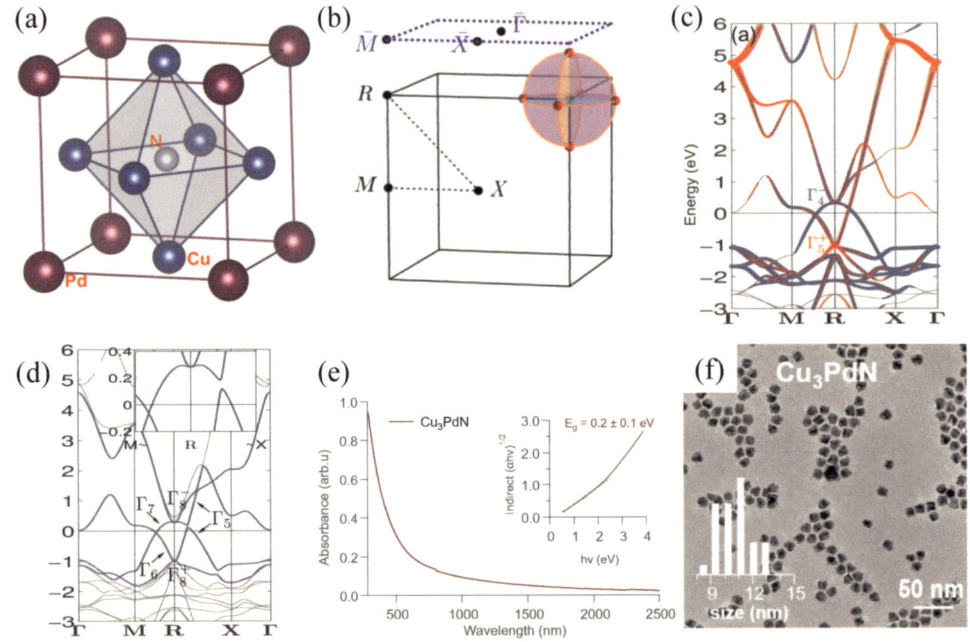

Figure 16.
(a) Crystal structure of anti-perovskite Cu_3PdN. (b) Bulk and projected (001) surface BZ, the orange rings and red points indicate the nodal line and Dirac points, respectively. Calculated electronic band structure of Cu_3PdN without (c) and with (d) SOC [32]. (e) UV–vis absorption spectrum, the corresponding Tauc plot is inserted on the right-up corner. (f) TEM image with statistical size inserted [33].

3.3 Copper-based topological semimetal

Topological semimetal is a kind of novel material whose CB and VB are connected at specific points or lines around E_F in the reciprocal space. In recent years, Cu_3PdN has been proposed as a Dirac semimetal, stabilized by the C_4 rotational crystal symmetry. As illustrated in **Figure 16(a)**, Cu_3PdN adopts an anti-perovskite crystal structure (space group $\boldsymbol{Pm\bar{3}m}$), with a nitrogen atom located at the cube's center, Cu atoms occupy the octahedral vertices, and Pd located at the corner site. **Figure 16(b)** shows the 3D bulk BZ and (001) projected surface BZ. For the calculated electronic band structure, in the absence of spin-orbital coupling (SOC), the VB and CB are dominated by the Pd 4d (blue) and Pd 5p (red) states, respectively. Accompanied by the energy band inversion around R point, the Dirac points near E_F are observed (**Figure 16(c)**). When introducing SOC, a tiny 62-meV gap is opened for the Dirac node along R-X direction (**Figure 16(d)**) [32]. Experimentally, the prepared Cu_3PdN nanocrystals show a cube morphology (**Figure 16(e)**) and a negligible band gap of 0.2 ± 0.1 eV (**Figure 16(f)**) [33].

Monolayer Cu_2Si is also theoretically and experimentally recognized as a topological semimetal. **Figure 17(a)** illustrates the crystal structure. As shown in **Figure 17(b)**, first-principle calculations reveal two hole-like bands (α, β) and one electron-like band (γ) in the energy band structure, which form closed contours on the Fermi surface (**Figure 17(c)**) with hexagonal, hexagram, and circular shapes, respectively. Besides, the γ-energy band intersects linearly with the α and β bands, forming two Dirac node loops (NL1, NL2) centered at the Γ point and protected by

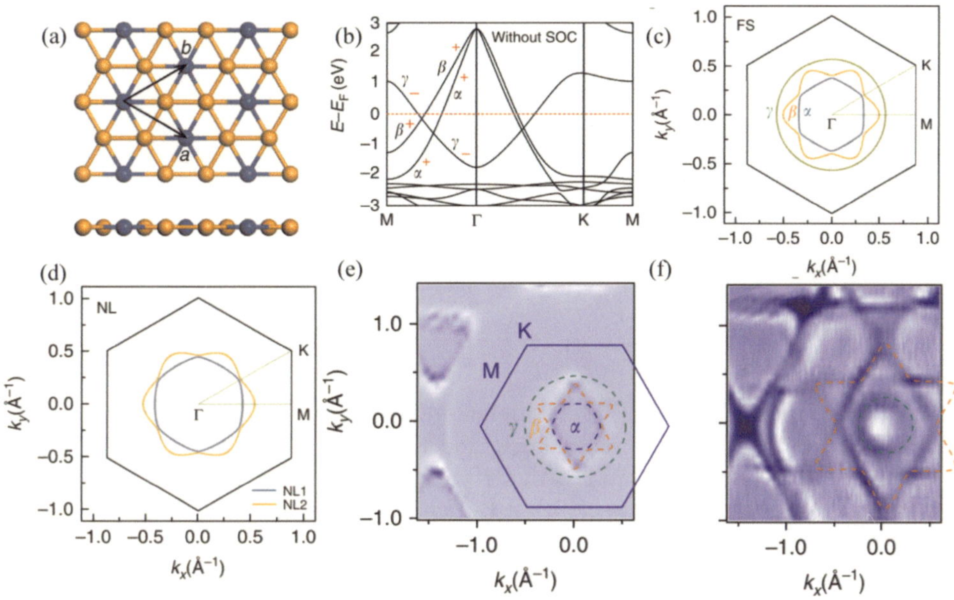

Figure 17.
(a) Top and side views of the atomic structure of monolayer Cu_2Si. (b) Calculated band structure, (c) Fermi surface, and (d) momentum distribution of the nodal loops (NL1, blue; NL2, orange) of Cu_2Si without SOC. The blue, orange, and green lines in (c-d) correspond to α, β, and γ bands in (b), respectively. (e-f) Experimental ARPES CECs at E_F and nodal point for monolayer $Cu_2Si/Cu(111)$ [34].

mirror inversion symmetry, as illustrated in **Figure 17(d)**. Experimentally, depositing Si atoms onto the Cu(111) surface through the molecular beam epitaxy (MBE) method can enable the formation of monolayer Cu_2Si films. High-resolution ARPES tests have also revealed the constant energy counters (CECs) at E_F (**Figure 17(e)**) and nodal point (**Figure 17(f)**), consistent with the theoretical prediction [34].

3.4 Copper-based superconductor

Superconductor is a kind of unique material that exhibits a sudden disappearance of electrical resistance and complete repulsion of magnetic fields, known as the Meissner effect, below the critical temperature (T_c) [35]. Under a superconducting state, the current within the superconductor can flow unimpededly without heat loss. This subsection provides several examples to introduce copper-based superconductors.

As shown in **Figure 18(a)**, polycrystalline samples of Ba_2CuO_{4-y} were synthesized using solid-state reaction at high pressure (~18 GPa) and high temperature (~ 1000°C). This sample exhibits an obvious superconducting transition at about 73 K in both zero-field-cooled (ZFC) and field-cooled (FC) modes (**Figure 18(b)**). This enhanced T_c than the isostructural La_2CuO_4-based counterparts can be attributed to the compressed octahedron structure, in which the $3d3z^2$-r^2 orbitals are lifted above the $3dx^2$-y^2 orbitals (**Figure 18(c)**) [36].

Crystalline $[Cu_3(C_6S_6)]_n$ (Cu-BHT) film is another kind of superconductor, and it can be synthesized through liquid-liquid interfacial reaction. **Figure 19(a–c)** present the atomic structure and high-resolution TEM images of Cu-BHT. Simply speaking,

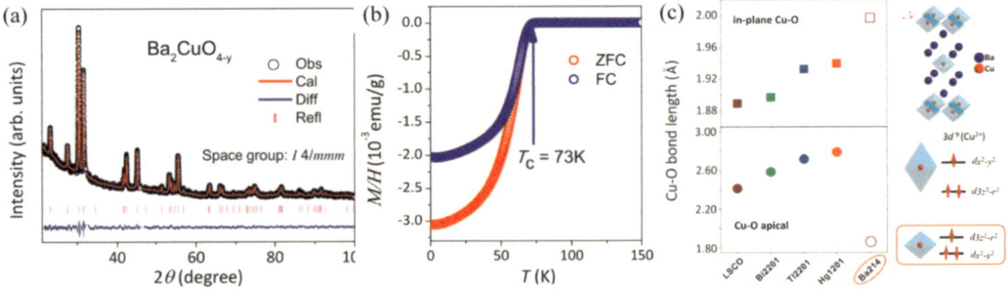

Figure 18.
(a) Typical XRD pattern, (b) magnetization, and (c) Cu-O bond lengths of Ba_2CuO_{4-y} *superconductor [36].*

Cu-BHT is composed of stacked π-d conjugated 2D nanosheets with Cu^{2+} ions and BHT ligands, and forming the famous kagome structure, that is, woven arrangement of six corner-shared triangles and form sixfold symmetry of structure composed of hexagons and triangles. In **Figure 19(d)**, the in-plane resistivity is plotted against temperature, revealing a sharp superconducting transition at around 0.25–0.3 K. Furthermore, the origination of the unconventional superconductivity of Cu-BHT is attributed to the strong electron correlations, due to T_c/T_F of Cu-BHT is 0.025 and belongs to the range of strongly correlated superconductors [37].

4. Application in energy conversion and catalysis

Over the past few decades, copper and its compounds have been applied in many areas of energy conversion and catalysis, including electrocatalysis of hydrogen evolution, oxygen reduction reaction, carbon dioxide reduction, and so on. Moreover, they contribute significantly to photocatalysis, enabling processes such as water decomposition and wastewater treatment. Additionally, in oxidation reaction catalysis, they play a pivotal role in oxidizing and utilizing organic compounds to produce environmentally friendly products. These applications are significantly important for improving energy utilization and mitigating environmental pollution. This section will briefly introduce the typical application of Cu and its compounds in energy conversion and catalysis.

4.1 Plasmonic catalysis

Plasmonic catalysis usually utilizes the plasmonic effect-induced hot electrons of metal and its compounds to enhance the catalytic reactions. The core principle involves the emergence of localized surface plasmon resonance (LSPR) upon light irradiation with a specific wavelength, which induces oscillations of free electrons on material surfaces and generates a strong electric field. This strong electric field produced by LSPR can interact with adjacent reactants or catalysts, leading to alterations in their chemical behavior. Additionally, LSPR can generate lots of charge carriers and induce photothermal effects [38]. Moreover, the plasmonic property can be tuned by the inherent characteristics, dimensions, and morphology of the materials [39].

Au, Ag, and Cu are the most commonly employed plasmonic materials, and Cu owns the advantages of being non-precious, cost-effective, and readily available. Furthermore, Cu nanoparticles (Cu NPs) exhibit superior LSPR effects within the

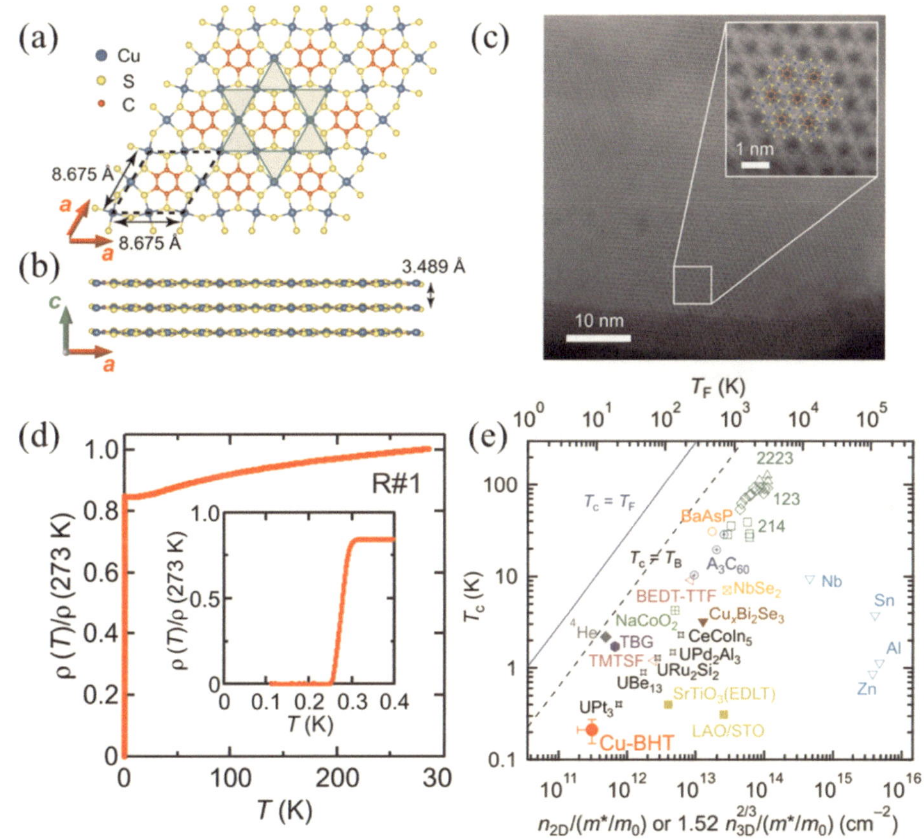

Figure 19.
(a–b) Top-view and side-view of the crystal structure of Cu-BHT, the triangular color blocks indicate the kagome structure. (c) HRTEM image. (d) Temperature dependence of the in-plane resistivity. (e) Uemura plot of T_c *against the effective superfluid density [37].*

UV-visible and even near-infrared range, making them broadly applied in water splitting, CO_2 reduction, and so on. As shown in **Figure 20**, the experimentally synthesized Cu NPs *via* photoreduction show obvious SPR absorption around 600 nm, and the photocatalytic activity can be optimized through adjusting the precursor Cu^{2+} amounts. As a result, the maximum hydrogen (H_2) and oxygen (O_2) evolution rates from water splitting reach 9.02 and 4.48 μmol·h^{-1}, respectively [40].

Li *et al*. reported that the hybridization of plasmonic Cu NPs onto TiO_2 to form *x* wt.%-Cu/TiO_2 can consistently increase the light absorption between 400 and 750 nm (**Figure 21(a)**). Besides, the decreased photoluminescence (PL) spectra (**Figure 21(b)**) inform the enhanced photogenerated carrier separation and reach the best state for 0.13 wt.%-Cu/TiO_2. Furthermore, the introduction of Cu NPs markedly improves the photocatalytic hydrogen production rate from 666.42 μmol·g^{-1}·h^{-1} for pure TiO_2 to 2853.53 μmol·g^{-1}·h^{-1} for 0.13 wt.%-Cu/TiO_2 (**Figure 21(c)**). This heightened efficiency is attributed to the plasmonic nature of Cu and the resultant photothermal effect [41].

Wang *et al*. reported that load Cu onto ZnO support to form Cu/ZnO catalysts can effectively improve the visible-light absorption with the peak centered at 584.3 nm (**Figure 21(d)**). Moreover, the methanol production rate from CO_2 reduction can be

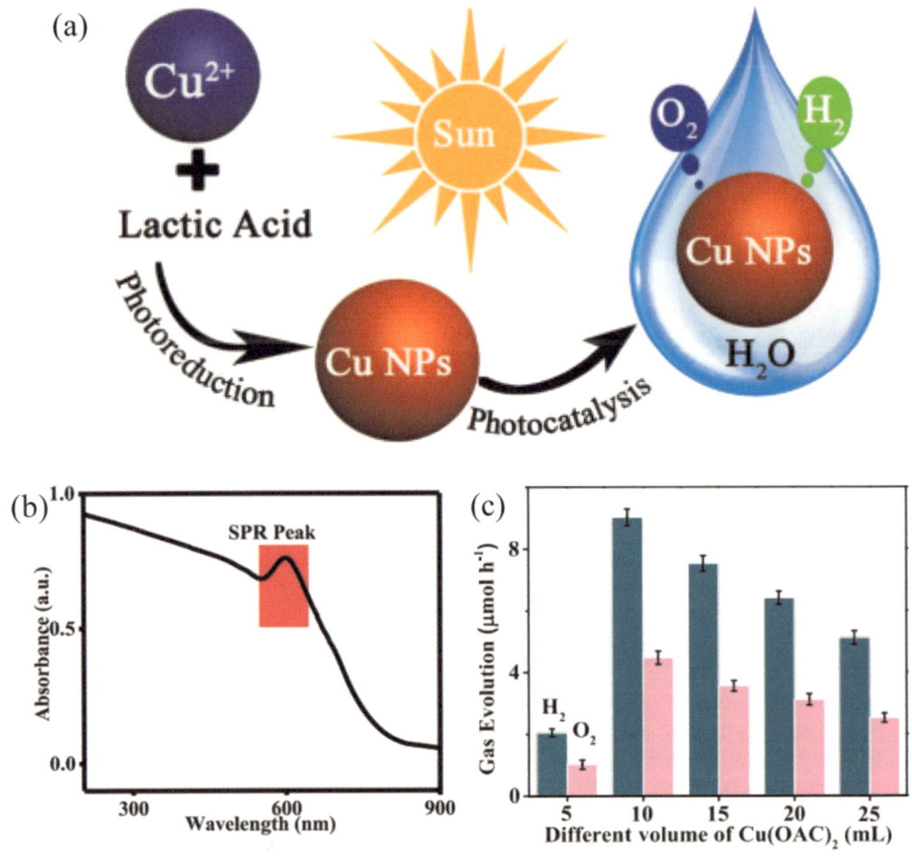

Figure 20.
(a) Schematic illustration of the synthesis and photocatalysis of Cu NPs. (b) Absorption spectrum. (c) The photocatalytic gas evolution from water with the Cu NPs prepared with different volumes of precursor [40].

improved from 1.38 μmol·g^{-1}·min^{-1} under dark to 2.13 μmol·g^{-1}·min^{-1} under visible-light irradiation (**Figure 21(e)**). Concurrently, the apparent activation energy is decreased by about 40% from 82.4 to 49.4 kJ·mol^{-1} (**Figure 21(f)**). These results indicate that the SPR-induced hot electrons of Cu NPs can effectively assist the activation of reaction intermediates during methanol synthesis [42].

4.2 Hydrogen evolution reaction (HER)

HER is an important half-reaction for (photo)electrocatalytic water splitting. For the standard electrolytic cell, it is composed of cathode, anode, and electrolyte. HER normally occurs at cathode. Under the external electric field, electrons can move to the cathode surface and participate in the reduction reaction to generate H_2. Besides, the electrolyte's acid-base property can influence the HER reaction route. For acidic electrolytes, the reaction can be expressed as $2H^+ + 2e^- \rightarrow H_2$. For alkaline electrolytes, it follows: $2H_2O + 2e^- \rightarrow H_2 + 2OH^-$ [43–44].

Cu and its compounds are widely employed as effective HER catalysts. For example, incorporating Cu_2S into CdZnS (CZS) to form snowflake-shaped 2% Cu_2S/CZS

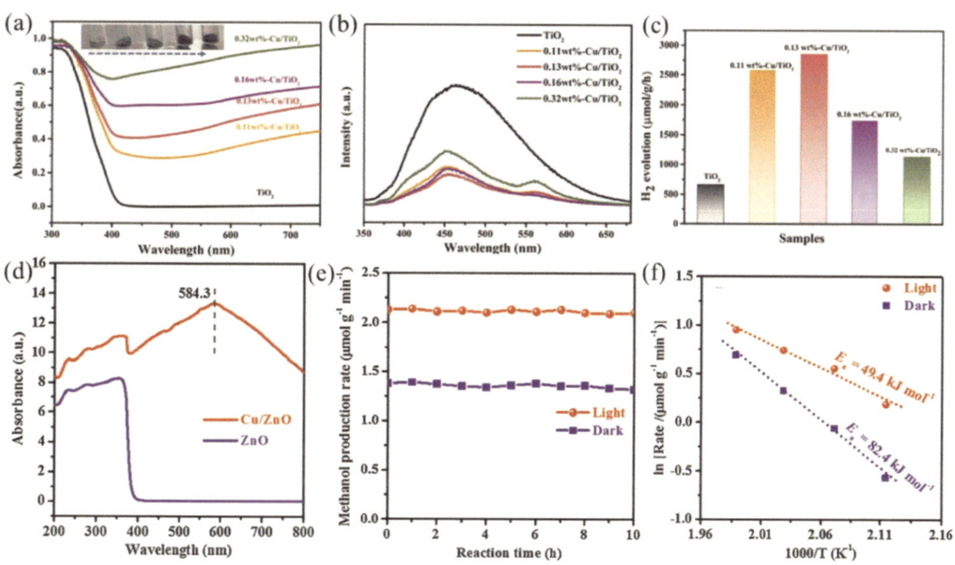

Figure 21.
(a) UV–vis DRS absorption spectra. (b) PL emission spectra. (c) H_2 evolution rate over x wt%-Cu/TiO_2 [41]. (d) UV–vis absorption spectra of Cu/ZnO catalyst and ZnO support. (e) Stability test of methanol production rate from CO_2 reduction over the Cu/ZnO catalyst at 220°C with and without visible light irradiation. (f) Arrhenius plots under dark and light conditions [42].

heterojunction (**Figure 22(a)**) can not only enhance the light absorption property (**Figure 22(b)**), but also improve the hydrogen evolution within 5 hours from 92.5 μmol for pristine CZS to 295.2 μmol for 2% Cu_2S/CZS (**Figure 22(c)**), due to the reduced agglomeration of granular CZS by snowflake-shaped structure of Cu_2S. Additionally, 2% Cu_2S/CZS shows great HER stability (**Figure 22(d)**) [45]. Similar experiments reveal that the deposition of Cu NPs can further facilitate the separation of photogenerated carriers in $TiO_2/CuInS_2$ (**Figure 22(e)**), and improve the HER activity even when $TiO_2/CuInS_2$ reaches a saturated state (**Figure 22(f)**) [46].

4.3 Oxygen evolution reaction (OER)

OER is another half-reaction for PEC water splitting, which happens at the anode. Similar to HER, the acid-base property of electrolytes affects the OER electrochemical reactions. Under acidic conditions, it follows $2H_2O \rightarrow 4H^+ + O_2 + 4e^-$. Under alkaline conditions, OER proceeds as $4OH^- \rightarrow 2H_2O + O_2 + 4e^-$. For OER, efficient catalysts are crucial to lower the activation energy and enhance the efficiency of electrochemical conversion, necessitating dedicated research for efficient OER catalysts to be crucial [47].

Cu and its compounds are widely used catalysts in OER reactions. As shown in **Figure 23**, CuO is composited with $CoMn_2O_4$ (CMO), forming CuO-CMO nanostructures, to enhance the OER performance of CMO. By optimizing the CuO concentration, CuO-CMO-16% demonstrates superior performance, with an overpotential of 277 mV at 10 mA·cm^{-2} (**Figure 23(a)**), the lowest Tafel slope of

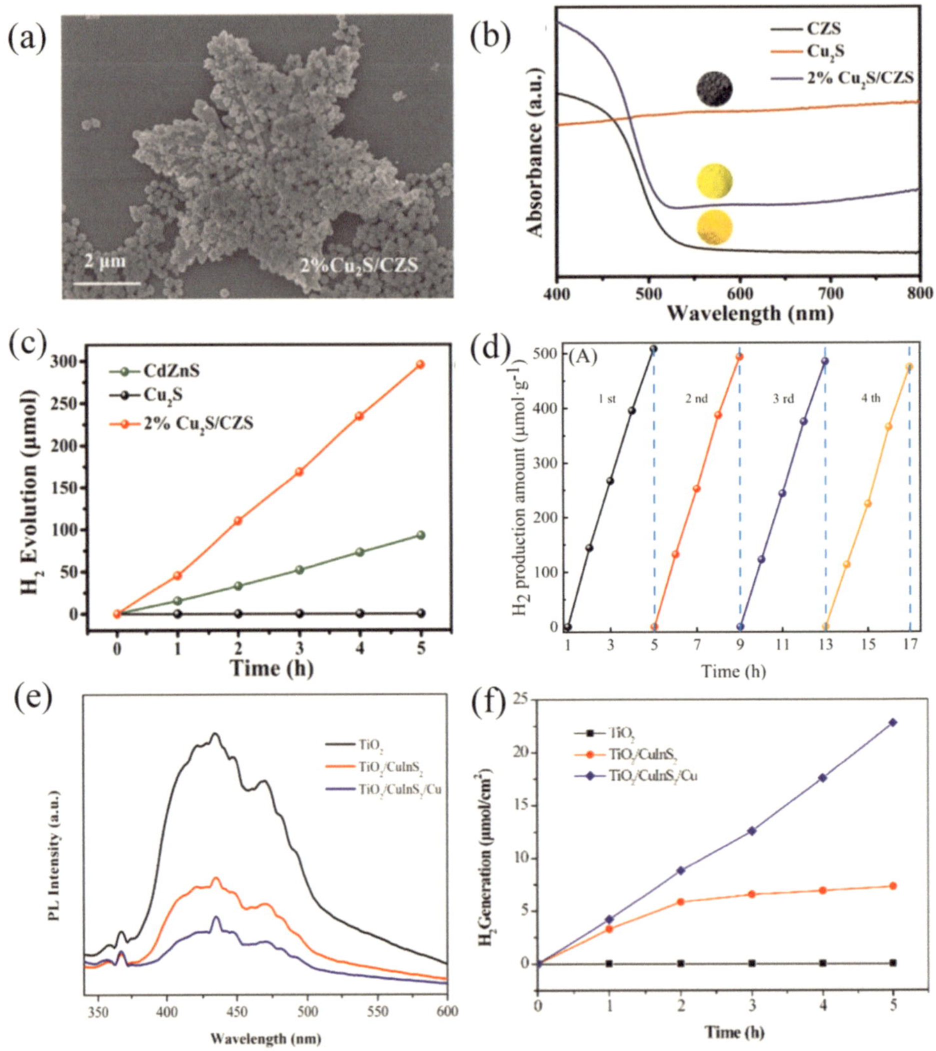

Figure 22.
(a) SEM image of 2%Cu_2S/CZS (CdZnS). (b) UV–vis DRS and (c) H_2 evolution of CZS, Cu_2S and 2%Cu_2S/CZS. (d) Cycling H_2 generation experiment [45]. (e) PL spectra and (f) H_2 generation of TiO_2, $TiO_2/CuInS_2$, and $TiO_2/CuInS_2/Cu$ under >420 nm visible light [46].

43 mV dec^{-1} (**Figure 23(b)**), larger electrochemical active surface area (ESCA = C_{dl}/C_{sp}, **Figure 23(c–e)**), and good long-term stability (**Figure 23(f)**) [48].

Porous Cu-CoP (Cu_1Co_xP) nanoplates were employed as OER catalysts, and $Cu_1Co_{10}P$ with a molar ratio of Cu to Co to be 1:10 shows the highest current density (**Figure 24(a)**) and low overpotential of 252 mV at 10 mA cm^{-2} (**Figure 24(c)**). Besides, $Cu_1Co_{10}P$ exhibits a low Tafel slope (89.1 mV dec^{-1}, (**Figure 24(b)**)), largest active surface area (37.21 mF cm^{-2}, (**Figure 24(d)**)), the lowest impedance (**Figure 24(e)**). Moreover, $Cu_1Co_{10}P$ shows superior stability under alkaline conditions at 1.48 V vs. RHE (reversible hydrogen electrode) (**Figure 24(f)**). The improvement in OER performance can be attributed to Cu doping and morphology modulation of porous structure that induce more surface electronic states and rapid electron transfer [49].

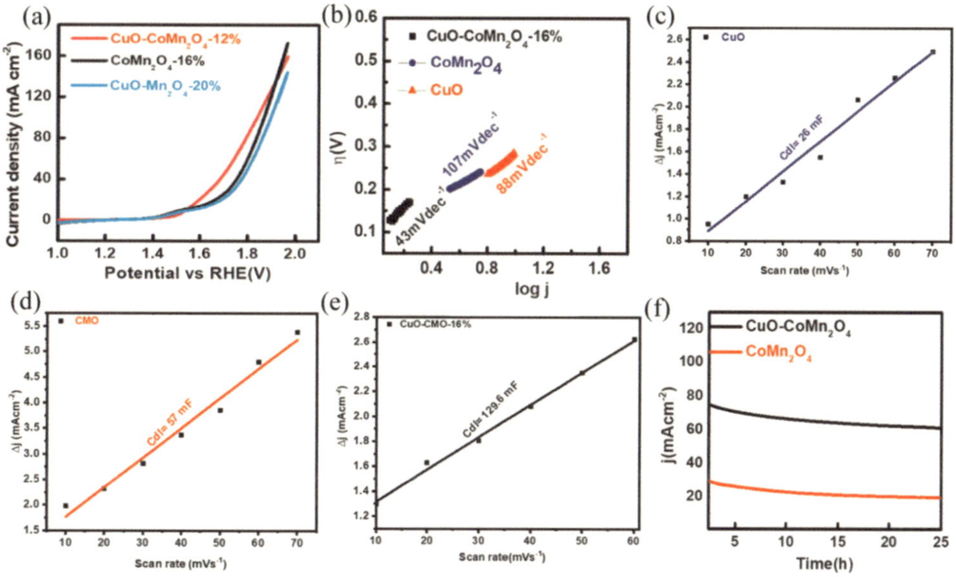

Figure 23.
(a) LSV curve for CuO-CMO-x with varying composition of CuO. (b) Tafel plot analysis. (c-e) C_{dl} plots for CMO, CuO, and CuO-CMO-16%. (f) Long-term stability test for CMO and CuO-CMO-16% [48].

4.4 Oxygen reduction reaction (ORR)

ORR is a crucial chemical reaction in fuel cells and metal-air batteries, and garnering extensive research attention in recent years. ORR involves the reduction of oxygen (O_2) to water (H_2O) or hydroxide (OH^-). As depicted in **Figure 25**, ORR can proceed through two-electron and four-electron pathways. The crucial distinction between the two ORR pathways relies on the cleavage of the O—O bond. If the O—O bond is broken, the reaction follows the two-electron ORR pathway; otherwise, it proceeds through the four-electron ORR pathway [50]. Based on pH conditions, ORR can be further categorized into four reactions [51]:

In acidic conditions:

Four-electron reduction: $O_2 + 4H^+ + 4e^- \rightarrow 2H_2O$;

Two-electron reduction: $O_2 + 2H^+ + 2e^- \rightarrow H_2O_2$.

In alkaline conditions:

Four-electron reduction: $O_2 + 2H_2O + 4e^- \rightarrow 4OH^-$;

Two-electron reduction: $O_2 + H_2O + 2e^- \rightarrow HO_2^- + OH^-$.

Coupling Cu-$CuFe_2O_4$ nanohybrid with conductive carbon to form Cu-$CuFe_2O_4$/C has been utilized as a highly efficient electrocatalyst for the ORR. As shown in **Figure 26(a)**, the well-defined ORR peak with a current density of -1.49 mA·cm^{-1} at -0.33 V is observed for Cu-$CuFe_2O_4$/C, exhibiting sharper and more pronounced characteristics and more positive ORR onset potential than that of other materials. The results of linear-sweep voltammetry (LSV) in **Figure 26(b)** also indicate that the onset and half-wave potentials for Cu-$CuFe_2O_4$/C, that is, -0.139 and $-$ 0.39 V vs. Ag/AgCl, respectively, are more positive than others, suggesting a superior ORR efficiency of Cu-$CuFe_2O_4$/C, which can be attributed to the presence of metallic Cu and the reduced size of NPs on the carbon substrate. Furthermore, Koutecky-Levich

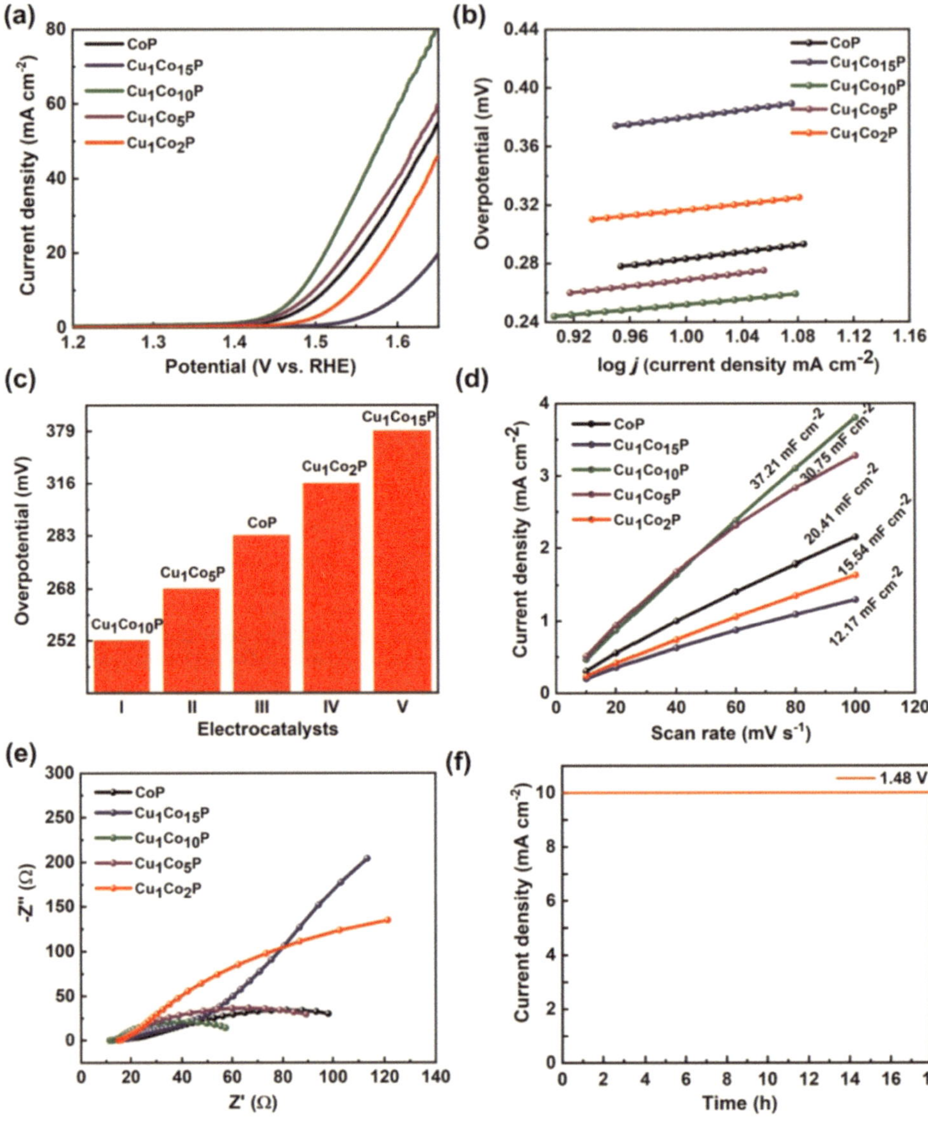

Figure 24.
(a) LSV curves, (b) Tafel plots, (c) overpotential at current density of 10 mA·cm^{-2}*, (d) scan-rate dependent current density, and (e) Nyquist plots of CoP and* Cu_1Co_xP *samples. (f) Long-term test for* $Cu_1Co_{10}P$ *[49].*

(K-L) analysis between −0.4 and −0.7 V indicates that the ORR on Cu-$CuFe_2O_4$/C proceeds a four-electron process **(Figure 26(c)–(d))** [52].

As depicted in **Figure 27**, the ultrathin flower-like 2D Cu/Cu_2O nanosheets grown through the low-temperature method in an ice/water mixed environment (273–278 K) demonstrate excellent ORR activity. CV curves exhibit a distinct reduction peak with an onset potential of 0.987 V and a half-wave potential of 0.921 V at 273 K (**Figure 27(a)**). First-principle calculations reveal that the abundant active Cu_2O stepped atoms can result in excellent low-temperature ORR activity. **Figure 27(b)** shows the calculated adsorption energies of OO^*, OOH^*, O^*, and OH^*. While the reaction steps of $Cu_2O(111)$ and $Cu_2O(220)$ exhibit a downhill trend, Cu(111) includes an uphill tendency and is not

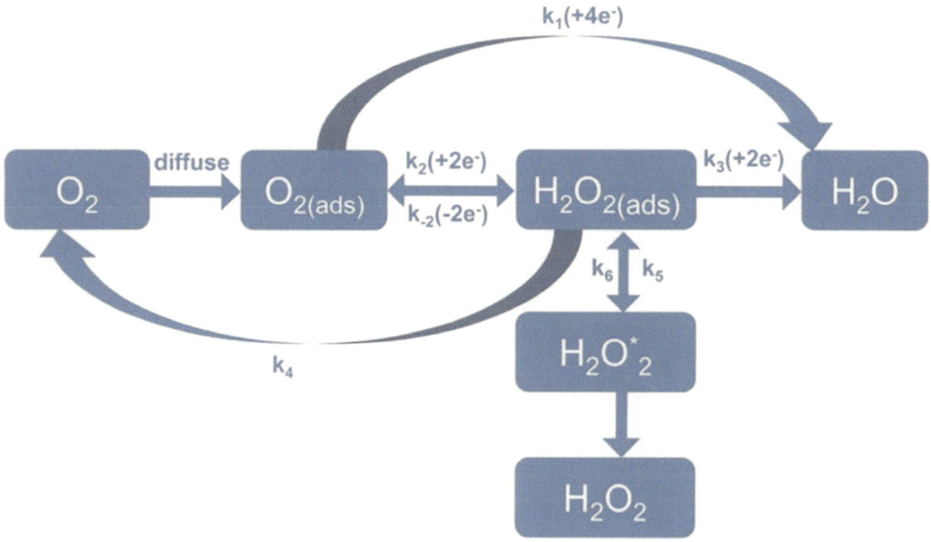

Figure 25.
Two-electron and four-electron pathways of ORR [51].

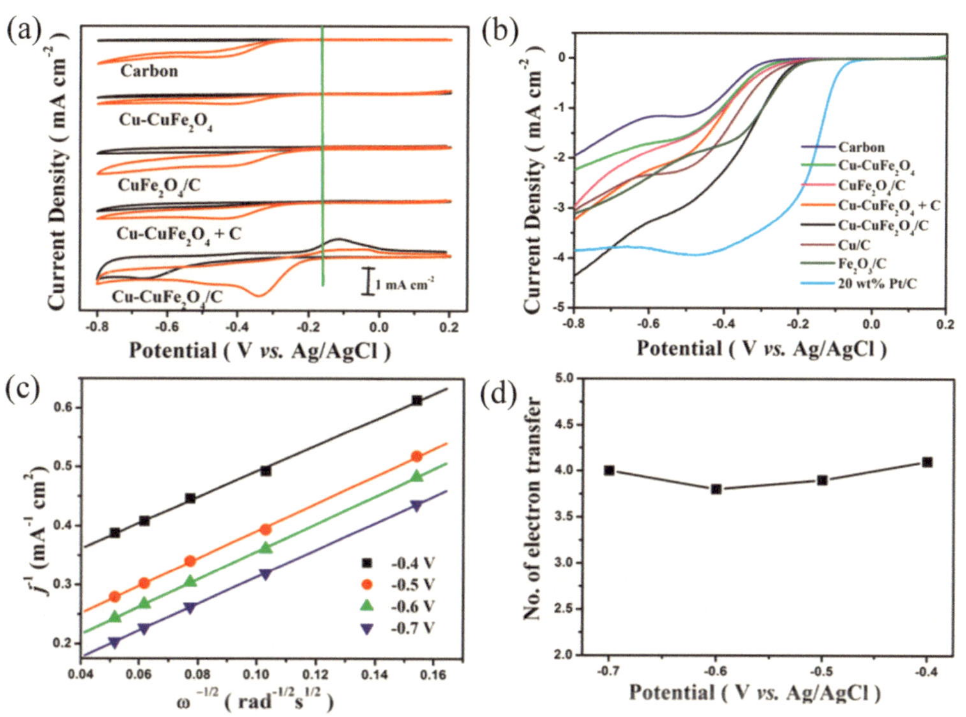

Figure 26.
(a) Cyclic voltammetry (CV) of several Cu-based materials in O_2- (red lines) and N_2-saturated (black lines) 0.1 M KOH electrolyte. (b) LSVs in an O_2-saturated 0.1 M KOH electrolyte at 1600 rpm. (c) K-L plots between −0.4 and − 0.7 V. (d) Plot of electron transfer vs. potential for Cu-$CuFe_2O_4$/C hybrid [52].

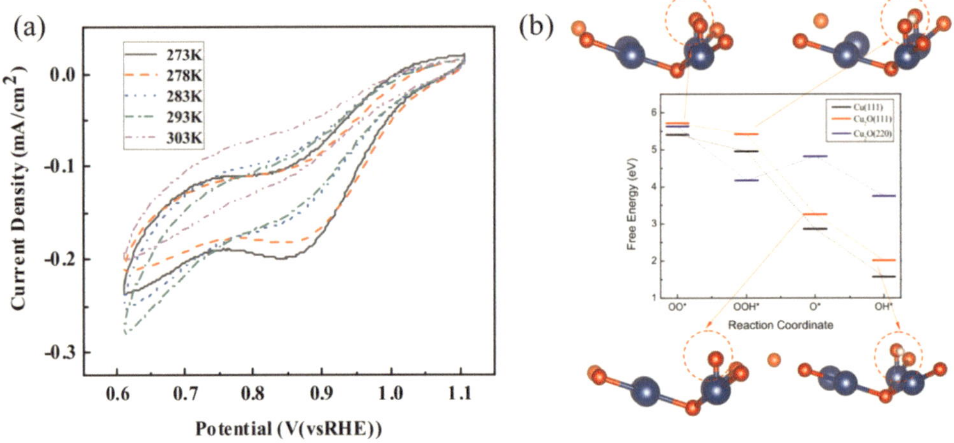

Figure 27.
(a) Low-temperature CV curves of 2D Cu/Cu_2O nanosheets in N_2- or O_2-saturated 0.1 Mol·L^{-1} KOH solution at a scan rate of 50 mV·s^{-1}. (b) First-principle calculated free energy for ORR over 2D Cu/Cu_2O nanosheets [53].

favorable for the ORR reaction. Furthermore, the difference in the step trends of Cu_2O(111), Cu_2O(220), and Cu(111) can be attributed to the d-band center theory, that is, the d-band centers of Cu_2O(220) (−0.65 eV) and Cu(111) (−0.78 eV) are more closer to E_F level (0.0 eV) than Cu(111) (−1.26 eV) [53].

4.5 Nitrogen reduction reaction (NRR)

NRR is a kind of reaction that can convert molecular nitrogen (N_2) into ammonia (NH_3) or other nitrogen-containing compounds, which are extensively applied in medicine, chemical engineering, and environmental sciences. It is a complex, multistep process involving several intermediates and electron transfer steps, such as adsorption, electron transfer, intermediate generation, and product release. Initially, the N_2 molecule is adsorbed onto the surface of the catalyst, followed by an electron transfer that facilitates the gradual breaking of the N≡N bond. Subsequently, the N_2 molecules are progressively transformed into various intermediates, such as N_2H^*, NH^*, NH_2^*, and ultimately yield NH_3 [54]. Besides, the six-electron process NRR is competitive with two-electron HER as their similarity in equilibrium potential in both acidic and alkaline electrolytes. Then, it is crucial to design a specific electrocatalyst that prefers to adsorb N atom than H atom, that is, suppress HER and enhance the selectivity of NRR [55].

In recent years, there has been a surge in the application of Cu-based catalysts in NRR. For instance, the *in situ* modification of Cu_2O with unsaturated Cu-MOF to form Cu-MOF/Cu_2O heterostructures shows an excellently boosted NH_3 production rate (7.16 mmol/m^2/h), even superior to the noble-metal Pt (**Figure 28(a)**). Besides, Cu-MOF/Cu_2O exhibits stable performance after five cycles (**Figure 28(b)**). The enhanced NRR performance can be attributed to the unsaturated Cu(II) sites and the porous structure of Cu-MOF, which can selectively absorb and activate N_2 molecules [56].

As shown in **Figure 29**, anchoring zero-valent Cu NPs onto reduced graphene oxide (Cu NPs-rGO) can serve as an efficient electrocatalyst for NRR [57]. It achieves an ammonia (NH_3) yield of 24.58 $\mu g \cdot h^{-1} \cdot mg^{-1}{}_{cat}$ and a Faraday efficiency (FE) of 15.32% at −0.4 V vs. RHE (**Figure 29(b)**). Furthermore, the production rate and FE show excellent cycling stability, the current density exhibits good long-term stability.

4.6 Carbon dioxide reduction reaction (CO_2RR)

CO_2RR is a process that converts CO_2 into renewable energy. The CO_2RR mechanism encompasses four key stages: adsorption, activation, electron transfer, and product desorption. Initially, CO_2 molecules undergo adsorption onto the electrode

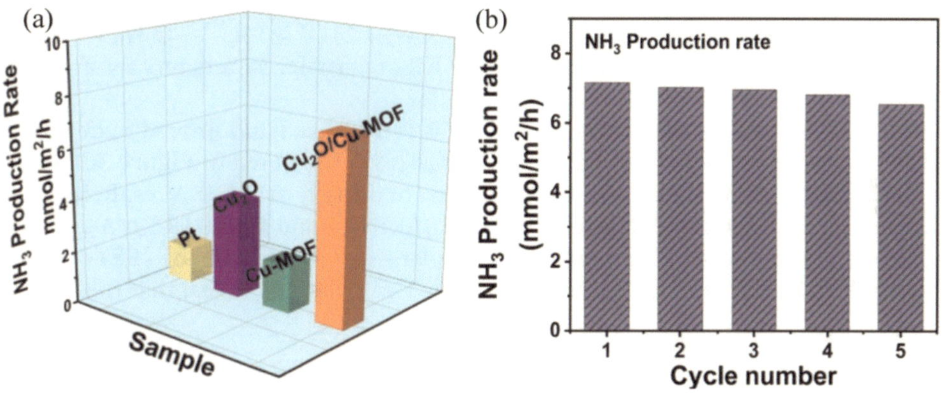

Figure 28.
(a) NH_3 production rate of Pt, Cu_2O, Cu-MOF and Cu-MOF/Cu_2O. (b) Recycling test of Cu-MOF/Cu_2O [56].

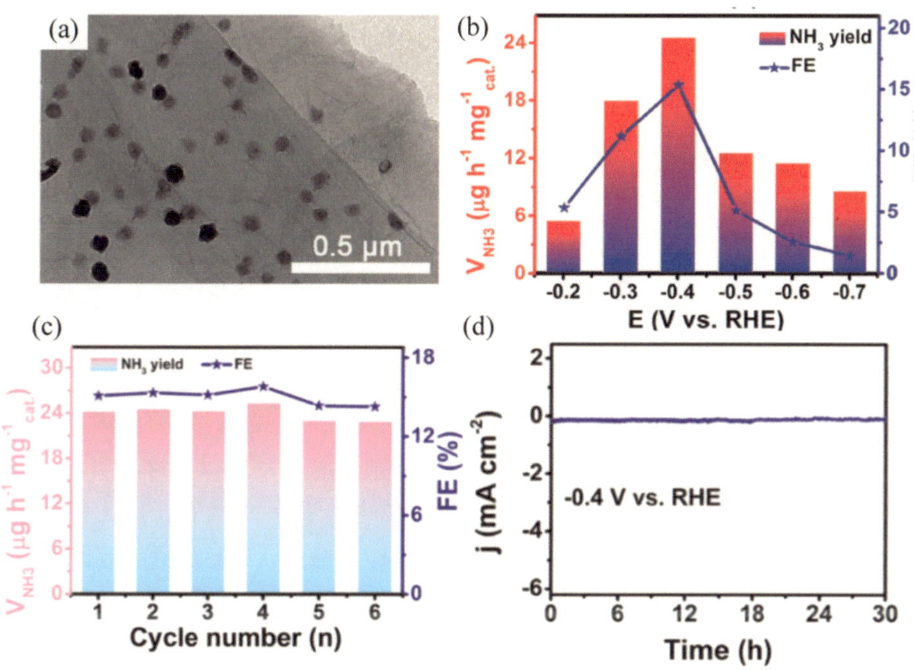

Figure 29.
(a) TEM image of the Cu NPs-rGO. (b) V_{NH_3} and FEs at various potentials. (c) Cycling test for NRR at −0.4 V. (d) Time-dependent current density curve for 30 h at −0.4 V [57].

surface. Subsequently, the adsorbed CO_2 undergoes activation to enhance its reactivity with electrons. The activated CO_2 molecules then react with the electrons on the electrode, accepting the electrons and forming a reduction intermediate. This intermediate subsequently undergoes further reactions to yield the final product. CO_2RR is a complex multi-electron process, and different reaction routes give rise to distinct intermediates and products, including single-carbon (C_1; HCOOH, CH_3OH, CH_4), two-carbon (C_2; C_2H_4, C_2H_6, C_2H_5OH), and even multi-carbon (C_n, $n \geq 3$) products. For example, a two-electron transfer results in the formation of formic acid (HCOOH) through the reaction $CO_2 + 2e^- + 2H^+ \rightarrow HCOOH$, a six-electron transfer yield methanol (CH_3OH) *via* $CO_2 + 6e^- + 6H^+ \rightarrow CH_3OH + H_2O$, and a twelve- and fourteen-electron process generate ethyl alcohol (C_2H_5OH) and ethane (C_2H_6) by $2CO_2 + 12e^- + 12H^+ \rightarrow C_2H_5OH + 3H_2O$ and $2CO_2 + 14H^+ + 14e^- \rightarrow C_2H_6 + 4H_2O$, respectively. The selectivity and yield of CO_2RR are crucial parameters for assessing the process's efficiency [58].

A nanodefective Cu nanosheet (n-CuNS) is synthesized and proved to be an efficient catalyst for CO_2RR to ethylene (C_2H_4) [59]. As shown in **Figure 30(a)**, n-CuNS exhibits superior ethylene partial current density at −1.48 V vs. RHE (66.5 mA·cm^{-2}) than nondefective CuNS (5 mA·cm^{-2}) and CuNP (8.5 mA·cm^{-2}). Besides, n-CuNS also shows an enhanced ethylene faradaic efficiency (FE) of 83.2% (**Figure 30(b)**), informing high selectivity. DFT calculations reveal that the

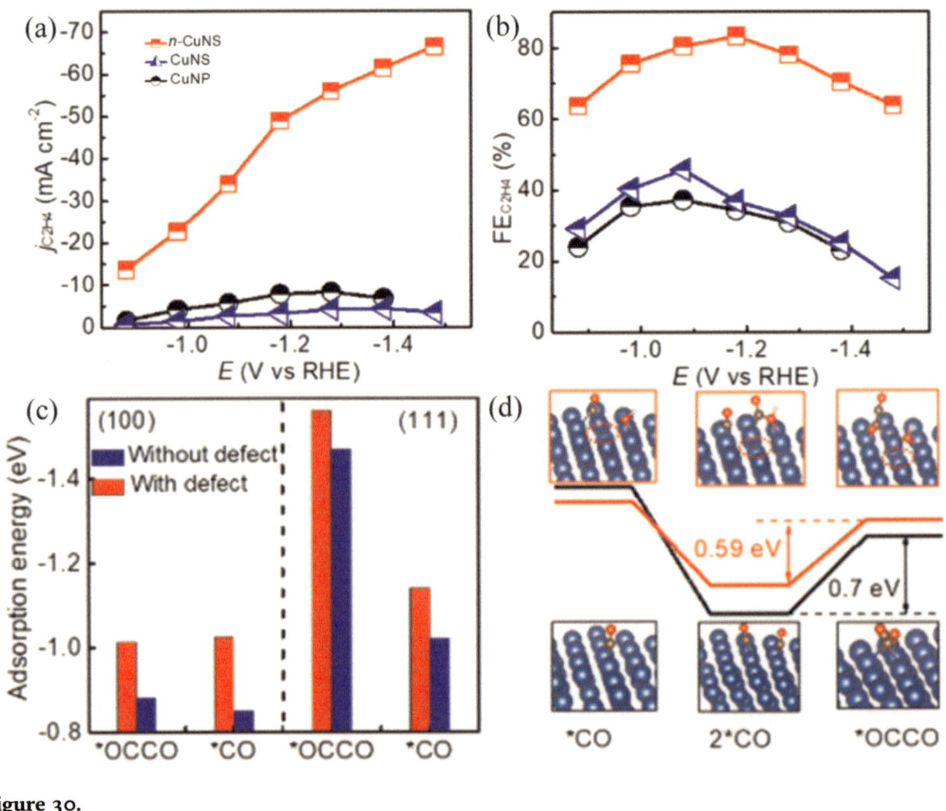

Figure 30.
(a) Ethylene (C_2H_4) partial current density and (b) FE from electrochemical CO_2RR at various potentials for different catalysts. (c) Comparison of adsorption energy of key intermediates on different facets. (d) Energy diagrams and geometries of CO dimerization on OH^- adsorbed and defective (red) and nondefective (black) Cu (111) of CuNS [59].

introduction of Cu defects can enhance the adsorption of key reaction intermediates $^*CO$, $^*OCCO$, and OH^- (**Figure 30(c)**). Besides, the defective Cu(111) surface of n-CuNS decreases the reaction barriers for 110 meV in the CO dimerization reaction ($^*CO + {}^*CO \rightarrow {}^*OCCO$), which is a rate-determining step for ethylene production (**Figure 30(d)**).

Choi *et al*. reported that constructing intimate atomic Cu-Ag interfaces on the surface of Cu nanowires through a two-step process (**Figure 31(a)**) can significantly enhance the methane (CH_4) selectivity of CO_2RR. As shown in **Figure 31(b)**, the CH_4 production from CO_2RR on Cu_9Ag_1NWs begins to increase at −1.12 V vs. RHE with FE_{CH4} of 63.29% ± 4.85%, and reaches a maximum FE_{CH4} of 72% at −1.17 V vs. RHE. Besides, Cu_9Ag_1NWs show obviously superior CO_2RR performance than CuNWs (**Figure 31(c)**), due to the synergistic effect of CO-Ag* and H-Cu* for enhanced CH_4 selectivity at low overpotential [60].

4.7 Single-atom catalysis (SAC)

SAC is an innovative catalytic technology in which the catalyst consists of individual atoms rather than clusters or multiple atoms. Typically, these catalysts anchor individual atoms to the surface of a solid carrier, ensuring both stability and high catalytic activity. A distinctive feature of SAC is its extremely high atomic utilization

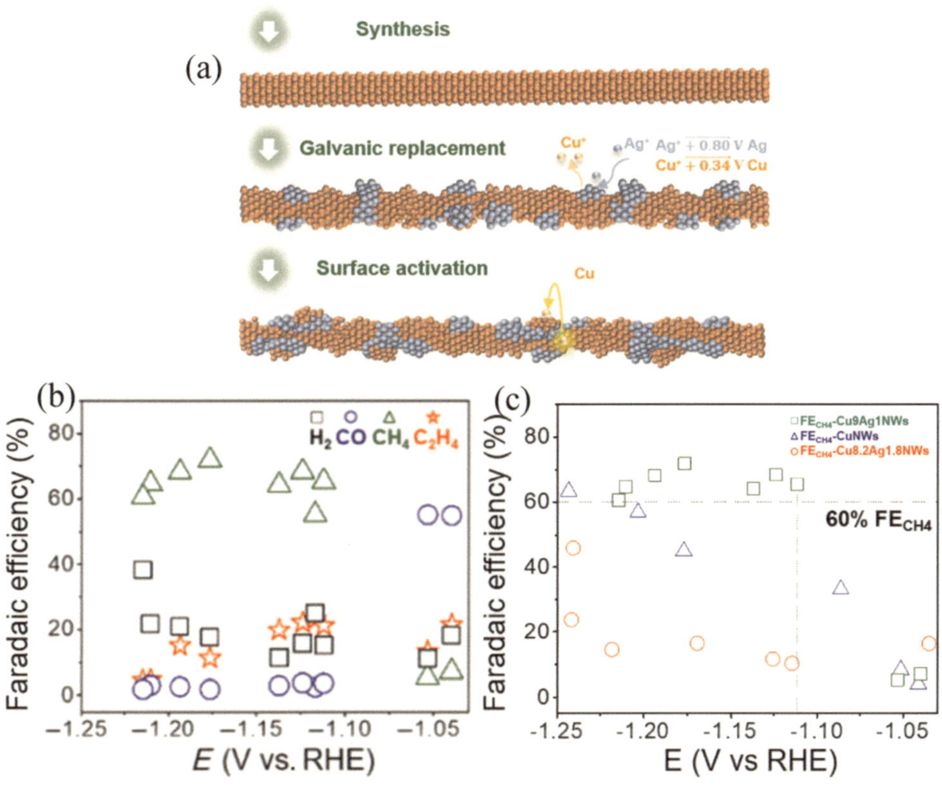

Figure 31.
(a) Schematic process of preparing CuAgNWs. (b) FEs of Cu_9Ag_1NWs in 0.1 M $KHCO_3$ at room temperature and atmospheric pressure. (c) Methane (CH_4) FEs of Cu_9Ag_1NWs, CuNWs and $Cu_{8.2}Ag_{1.8}NWs$ samples [60].

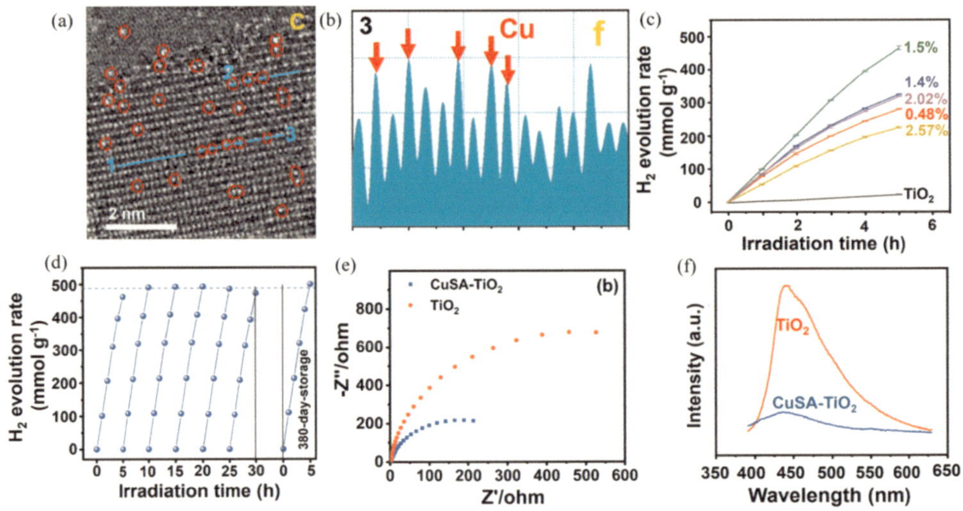

Figure 32.
(a) High-resolution HAADF STEM image of CuSA-TiO_2. (b) Line profile marked in (a). (c) Photocatalytic H_2 evolution rate of TiO_2 and TiO_2 loaded samples with different ratios of Cu SACs. (d) Cyclic photocatalytic activity of water splitting for 1.5 wt% CuSA-TiO_2. (e) ESI and (f) PL spectra of TiO_2 and CuSA-TiO_2 [62].

and density of active sites, stemming from the ability of each atom to act as an active site in the catalytic reaction. Moreover, the atomically dispersed nature allows SAC's electron clouds interact more efficiently with the reactant and substrate, further enhancing the catalytic activity. Due to its elevated activity, selectivity, and efficient atom utilization, SAC has become a rising star in catalysis research [61].

As shown in **Figure 32**, outstanding H_2 evolution performance is achieved through loading highly dispersed Cu single atoms onto TiO_2 and forming CuSA-TiO_2 [62]. The dispersion of atomic Cu is confirmed through the high-resolution high-angle annular dark-field (HAADF) STEM characterization (**Figure 32(a,b)**). The highest hydrogen yield occurred at a CuSA loading of 1.5 wt%, achieving an H_2 evolution rate of about 100 mmol·g^{-1}·h^{-1}, highlighting the pivotal role of CuSA loading (**Figure 32(c)**). Furthermore, it shows highly stable performance, even after 380-day storage in the lab (**Figure 32(d)**). Besides, the electrochemical impedance spectra (EIS) suggest that 1.5 wt% CuSA-TiO_2 exhibits lower impedance than pure TiO_2 (**Figure 32(e)**), more favorable for carrier transfer. The PL spectra in **Figure 32(f)** demonstrate that the photogenerated carriers of TiO_2 are effectively suppressed after CuSA loading, thereby promoting the separation of photogenerated carriers. These results imply that the anchoring of CuSA on TiO_2 can provide more active sites, enhance the photogenerated carrier separation and transfer, and promote the H_2 evolution.

Through a semi-transformed strategy, the Cu SAC is embedded into carbon dots to form Cu-CDs with the coordination of two N and two O. The synthesized N,O-coordinated Cu SAC (also CuN_2O_2) shows effective electrochemical CO_2RR activity to produce methane (CH_4), with the CH_4 partial current density of Cu-CDs 25 times higher than that of CuPc at −1.64 V vs. RHE, and the maximum turnover frequency (TOF) of Cu-CDs reaches a value of 2370 h^{-1}, suggesting a great CO_2RR activity (**Figure 33(a)**). Besides, the Cu-CDs catalyst shows stable performance of current density and high CH_4 Faradaic efficiency (78 ± 2%) (**Figure 33(b)**), revealing a remarkable selectivity of CO_2RR to CH_4. DFT calculations of the limiting potentials of

DOI: http://dx.doi.org/10.5772/intechopen.1004179

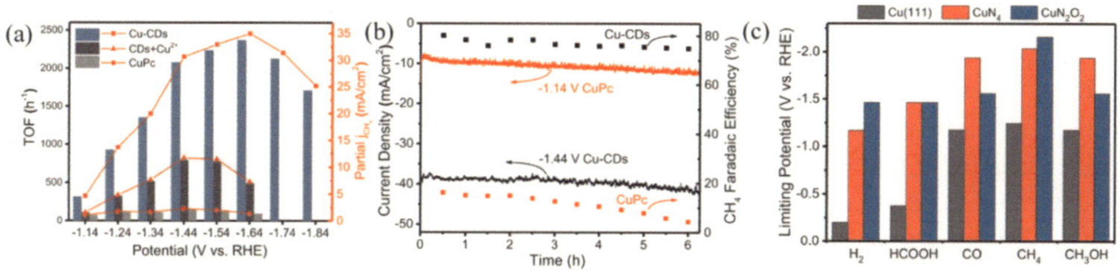

Figure 33.
(a) Partial CH_4 current density (right y-axis) and TOFs (left y-axis) of Cu-CDs, CDs + Cu^{2+}, and CuPc at different potentials. (b) Stability of current density and FE of Cu-CDs (black) and CuPc (red). (c) The limiting potentials of the products of CO_2RR and HER on the CuN_2O_2(Cu-CDs), CuN_4, and Cu(111) [63].

the products of CO_2RR and HER are depicted in **Figure 33(c)**, informing that the superior CH_4 selectivity of CuN_2O_2 (Cu-CDs) originates from the more negative potentials of CH_4 depart from other products for Cu-CDs, so that CH_4 can be exclusively produced with highest FE and less by-products [63].

5. Conclusions

This chapter provides a comprehensive introduction to the applications of copper and its compounds in energy conversion and catalysis. Based on the introduction of the crystal and electronic structure of bulk Cu and the representative low-index surfaces, several kinds of Cu-based materials with different band gaps are further illustrated, including semiconductors, semimetals, alloys, and superconductors. We find that the bonded elements and crystal structure can greatly influence the electronic band structure of the Cu-based compounds, and further induce various physical-chemistry properties, including light absorption property, conductivity, catalytic performance, and so on. At last, the typical application of Cu-based compounds in several important catalytic processes, utilizing solar energy and/or small quantity of electric energy, are introduced with typical examples. In general, Cu and Cu-based materials exhibit broad and effective applications in the critical energy conversion and catalysis fields, including plasmonic catalysis, HER, OER, ORR, NRR, CO_2RR, and SAC.

6. Perspectives

Despite significant progress in the field of energy conversion and catalysis using copper and its compounds have been achieved, there still have some various challenges and issues that demand attention. Firstly, how to design materials and further improve the activity, stability, and selectivity of specific reactions. Secondly, developing convenient theoretical and experimental methods for precise control of the atomic structure and electronic band structure of Cu-based compounds and the related heterojunction, as well as the number, location, and coordination environment of outer surface/inner layer active sites. Thirdly, an effective route for quantity production to meet the requirements of industrialization.

In the future, emphasis should be placed on catalytic activity, stability, selectivity, and quantity production, which are the keys to achieve industrialization. Initially, attention must be directed toward structural material design, involving alterations in specific surface area, pore size, and constructing heterostructure to enhance catalytic activity. Subsequently, stability can be heightened through surface modification utilizing corrosion-resistant and high-temperature-resistant materials. Furthermore, the establishment of stability in reaction intermediates *via* bond optimization between product and catalyst surface is crucial for controlling product selectivity. Lastly, artificial intelligence (AI)-based model study and data mining from historical database can be employed for structure design, performance prediction, and structure-relationship analysis, this should be quite helpful for shortening the development cycle of high-performance catalysts. Besides, efficient synthesis methods for fast and large-amount preparation must also be developed.

In summary, Cu and its compounds demonstrate substantial potential for applications in the realm of energy conversion and catalysis. Through sustained and in-depth research and exploration in future, Cu and its compounds could be believed to realize more potential and play a vital role in the field of energy conversion and catalysis.

Acknowledgements

Project is funded by the Talent Fund of Hubei University of Technology (grants BSQD2020112, BSQD2020107, BSQD2020104), Science and Technology Research Project of Education Department of Hubei Province (grant Q20201402), and the Outstanding Talent Foundation for Green Industry Leading Plan of HBUT (JCRC2021003).

Conflict of interest

The authors declare no conflict of interest.

Author details

Zhengwang Cheng[1,2*], Shengjia Li[1], Mei Wang[1,2] and Xinguo Ma[1,2]

1 School of Science, Hubei University of Technology, Wuhan, China

2 National "111 Research Center" Microelectronics and Integrated Circuits, Hubei University of Technology, Wuhan, China

*Address all correspondence to: zwcheng@hbut.edu.cn

References

[1] Chen GZ, Chen KJ, Fu JW, Liu M. Tracking dynamic evolution of catalytic active sites in photocatalytic CO_2 reduction by in situ time-resolved spectroscopy. Rare Metals. 2020;**39**: 607-609

[2] Elbaz AM, Mannaa O, Roberts WL. Flame flow field interaction in non-premixed CH_4/H_2 swirling flames. International Journal of Hydrogen Energy. 2021;**46**: 30494-30509

[3] Sun L, Han L, Huang J, Luo X, Li X. Single-atom catalysts for photocatalytic hydrogen evolution: A review. International Journal of Hydrogen Energy. 2022;**47**: 17583-17599

[4] IEA. Global Energy Review 2021. Paris: IEA; 2021. Available from: https://www.iea.org/reports/global-energy-review-2021

[5] Huang W, Dai J, Xiong L. Towards a sustainable energy future: Factors affecting solar-hydrogen energy production in China. Sustainable Energy Technologies and Assessments. 2022;**52**: 102059

[6] Durgalakshmi D, Ajay Rakkesh R, Rajendran S, Naushad M. Principles and mechanisms of green photocatalysis. In: Naushad M, Rajendran S, Lichtfouse E, editors. Green Photocatalysts. Cham: Springer International Publishing; 2020. pp. 1-24

[7] Rabaia MKH, Abdelkareem MA, Sayed ET, Elsaid K, Chae KJ, Wilberforce T, et al. Environmental impacts of solar energy systems: A review. Sciece of the Total Environment. 2021;**754**:141989

[8] Tembhare SP, Barai DP, Bhanvase BA. Performance evaluation of nanofluids in solar thermal and solar photovoltaic systems: A comprehensive review. Renewable and Sustainable Energy Reviews. 2022;**153**:111738

[9] Yu R, Wu G, Tan Z. Realization of high performance for PM6:Y6 based organic photovoltaic cells. Journal of Energy Chemistry. 2021;**61**:29-46

[10] Hu Z, Cheng Z, Gan N, Liu Z, Han C, Wang M, et al. Enhancing the photoelectrochemical performance of TiO_2 through decorating a topological insulator Bi_2Te_3 film and non-noble plasmonic Cu nanoparticles. Journal of Physical Chemistry C. 2022;**126**: 19047-19055

[11] Yue M, Lambert H, Pahon E, Roche R, Jemei S, Hissel D. Hydrogen energy systems: A critical review of technologies, applications, trends and challenges. Renewable and Sustainable Energy Reviews. 2021;**146**:111180

[12] Arsad AZ, Hannan MA, Al-Shetwi AQ, Mansur M, Muttaqi KM, Dong ZY, et al. Hydrogen energy storage integrated hybrid renewable energy systems: A review analysis for future research directions. International Journal of Hydrogen Energy. 2022;**47**: 17285-17312

[13] Burton NA, Padilla RV, Rose A, Habibullah H. Increasing the efficiency of hydrogen production from solar powered water electrolysis. Renewable and Sustainable Energy Reviews. 2021; **135**:110255

[14] Nowicki M, Wandelt K. Anion interaction with copper surfaces: General properties of metal surfaces. In:

Chiarotti G, Chiaradia P, editors. Physics of Solid Surfaces Subvolume B. Berlin: SpringerMaterials; 2018

[15] Valbuena MA, Walczak L, Martínez-Blanco J, Vobornik I, Segovia P, Michel EG. Lateral confinement effects of $\overline{M}$-point Tamm state in vicinal Cu (100) surfaces. Surface Science. 2014; **630**:144-152

[16] Jiang J, Tsirkin SS, Shimada K, Iwasawa H, Arita M, Anzai H, et al. Many-body interactions and Rashba splitting of the surface state on Cu(110). Physical Review B. 2014; **89**:085404

[17] Deng J, Xia B, Ma X, Chen H, Shan H, Zhai X, et al. Epitaxial growth of ultraflat stanene with topological band inversion. Nature Materials. 2018;**17**: 1081-1086

[18] Yin R, Xia J, Jiang B, Guo H. Theoretical insights into structure sensitivity in formate decomposition dynamics on Cu surfaces. ACS Catalysis. 2023;**13**:14103-14111

[19] Huang Y, Handoko AD, Hirunsit P, Yeo BS. Electrochemical reduction of CO_2 using copper single-crystal surfaces: Effects of CO^* coverage on the selective formation of ethylene. ACS Catalysis. 2017;**7**:1749-1756

[20] Spencer JA, Mock AL, Jacobs AG, Schubert M, Zhang YH, Tadjer MJ. A review of band structure and material properties of transparent conducting and semiconducting oxides: Ga_2O_3, Al_2O_3, In_2O_3, ZnO, SnO_2, CdO, NiO, CuO, and Sc_2O_3. Applied Physics Reviews. 2022;**9**: 011315

[21] Murali DS, Kumar S, Choudhary RJ, Wadikar AD, Jain MK, Subrahmanyam A. Synthesis of Cu_2O from CuO thin films: Optical and electrical properties. AIP Advances. 2015;**5**:047143

[22] Radhakrishnan AA, Beena BB. Structural and optical absorption analysis of CuO nanoparticles. Indian Journal of Advances in Chemical Science. 2014;**2**:158-161

[23] Balık M, Bulut V, Erdogan IY. Optical, structural and phase transition properties of Cu_2O, CuO and Cu_2O/CuO: Their photoelectrochemical sensor applications. International Journal of Hydrogen Energy. 2019;**44**:18744-18755

[24] Lee YS, Winkler MT, Siah SC, Brandt R, Buonassisi T. Hall mobility of cuprous oxide thin films deposited by reactive direct-current magnetron sputtering. Applied Physics Letters. 2011;**98**:193115

[25] Paredes P, Rauwel E, Rauwel P. Surveying the synthesis, optical properties and photocatalytic activity of Cu_3N nanomaterials. Nanomaterials. 2022;**12**:2218

[26] Mukhopadhyay AK, Momin MA, Roy A, Das SC, Majumdar A. Optical and electronic structural properties of Cu_3N thin films: A first-principles study (LDA + U). ACS Omega. 2020;**5**:31918-31924

[27] Wang LC, Liu BH, Su CY, Liu WS, Kei CC, Wang KW, et al. Electronic band structure and electrocatalytic performance of Cu_3N nanocrystals. ACS Applied Nano Materials. 2018;**1**: 3673-3681

[28] Ji AL, Huang R, Du Y, Li CR, Wang YQ, Cao ZX. Growth of stoichiometric Cu_3N thin films by reactive magnetron sputtering. Journal of Crystal Growth. 2006;**295**:79-83

[29] Gao MY, Yang C, Zhang QB, Yu YW, Hua YX, Li Y, et al.

Electrochemical fabrication of porous Ni-Cu alloy nanosheets with high catalytic activity for hydrogen evolution. Electrochimica Acta. 2016; **215**:609-616

[30] Jin Y, Chen F. Facile preparation of Ag-Cu bifunctional electrocatalysts for zinc-air batteries. Electrochimica Acta. 2015;**158**:437-445

[31] Wang Y, Zhu R, Wang Z, Huang Y, Li Z. Cu induced formation of dendritic CoFeCu ternary alloys on Ni foam for efficient oxygen evolution reaction. Journal of Alloys and Compounds. 2021; **880**:160523

[32] Yu R, Weng H, Fang Z, Dai X, Hu X. Topological node-line semimetal and Dirac semimetal state in antiperovskite Cu_3PdN. Physical Review Letters. 2015; **115**:036807

[33] Parvizian M, Balsa AD, Pokratath R, Kalha C, Lee S, Eynden DV, et al. The chemistry of Cu_3N and Cu_3PdN nanocrystals. Angewandte Chemie, International Edition. 2022;**61**: e202207013

[34] Feng B, Fu B, Kasamatsu S, Ito S, Cheng P, Liu CC, et al. Experimental realization of two-dimensional Dirac nodal line fermions in monolayer Cu_2Si. Nature Communications. 2017; **8**:1007

[35] Bussmann-Holder A, Keller H. High-temperature superconductors: Underlying physics and applications. Zeitschrift für Naturforschung. Teil B. 2020;**75**:3-14

[36] Li WM, Zhao JF, Cao LP, Jin CQ. Superconductivity in a unique type of copper oxide. Proceedings of the National Academy of Sciences of the United States of America. 2019;**116**: 12156-12160

[37] Takenaka T, Ishihara K, Roppongi M, Miao Y, Mizukami Y, Makita T, et al. Strongly correlated superconductivity in a copper-based metal-organic framework with a perfect kagome lattice. Science Advances. 2021; **7**:eabf3996

[38] Li S, Miao P, Zhang Y, Wu J, Zhang B, Du Y, et al. Recent advances in plasmonic nanostructures for enhanced photocatalysis and electrocatalysis. Advanced Materials. 2020;**33**:2000086

[39] Gellé A, Jin T, de la Garza L, Price GD, Besteiro LV, Moores A. Applications of plasmon-enhanced nanocatalysis to organic transformations. Chemical Reviews. 2019;**120**:986-1041

[40] Zhang P, Liu H, Li X. Photo-reduction synthesis of Cu nanoparticles as plasmon-driven non-semiconductor photocatalyst for overall water splitting. Applied Surface Science. 2021;**535**: 147720

[41] Li J, Hatami M, Huang Y, Luo B, Jing D, Ma L. Efficient photothermal catalytic hydrogen production via plasma-induced photothermal effect of Cu/TiO_2 nanoparticles. International Journal of Hydrogen Energy. 2023;**48**: 6336-6345

[42] Wang Z, Song H, Pang H, Ning Y, Dao TD, Wang Z, et al. Photo-assisted methanol synthesis via CO_2 reduction under ambient pressure over plasmonic Cu/ZnO catalysts. Applied Catalysis B: Environmental. 2019;**250**: 10-16

[43] Zhou F, Zhou Y, Liu GG, Wang CT, Wang J. Recent advances in nanostructured electrocatalysts for hydrogen evolution reaction. Rare Metals. 2021;**40**:3375-3405

[44] Wang J, Yue X, Yang Y, Sirisomboonchai S, Wang P, Ma X, et al. Earth-abundant transition-metal-based bifunctional catalysts for overall electrochemical water splitting: A review. Journal of Alloys and Compounds. 2020;**819**:153346

[45] Wang G, Quan Y, Yang K, Jin Z. EDA-assisted synthesis of multifunctional snowflake-Cu_2S/CdZnS S-scheme heterojunction for improved the photocatalytic hydrogen evolution. Journal of Materials Science and Technology. 2022; **121**:28-39

[46] Cheng Z, Hu Z, Ma X, Wang M, Gan N, Pan M. Enhancing the visible light photoelectrochemical water splitting of TiO_2 photoanode via a p–n heterojunction and the plasmonic effect. Journal of Physical Chemistry C. 2022; **126**:11510-11517

[47] Kannimuthu K, Sangeetha K, Sankar SS, Karmakar A, Madhu R, Kundu S. Investigation on nanostructured Cu-based electrocatalysts for improvising water splitting: A review. Inorganic Chemistry Frontiers. 2021;**8**:234-272

[48] Ashiq MF, Abid AG, Jabbour K, Fawy KF, Alzahrani HA, Ehsan MF. Increasing electrocatalytic efficiency of CuO/$CoMn_2O_4$ nanocomposite for oxygen evolution reaction. Ceramics International. 2023;**49**:28071-28079

[49] Thiyagarajan D, Gao M, Sun L, Dong X, Zheng D, Wahab MA, et al. Nanoarchitectured porous Cu-CoP nanoplates as electrocatalysts for efficient oxygen evolution reaction. Chemical Engineering Journal. 2022;**432**: 134303

[50] Wang T, Chutia A, Brett DJL, Shearing PR, He G, Chai G, et al. Palladium alloys used as electrocatalysts for the oxygen reduction reaction. Energy & Environmental Science. 2021; **14**:2639-2669

[51] Tang L, Xu Q, Zhang Y, Chen W, Wu M. MOF/PCP-based electrocatalysts for the oxygen reduction reaction. Electrochem Energy R. 2021;**5**:32-81

[52] Borah BJ, Yamada Y, Bharali P. Unravelling the role of metallic Cu in Cu-$CuFe_2O_4$/C nanohybrid for enhanced oxygen reduction electrocatalysis. ACS Applied Energy Materials. 2020;**3**: 3488-3496

[53] Li Z, Zhang Y, Yu H, Zhao H. Low temperature growth of two-dimensional (2D) Cu/Cu_2O nanosheets under ice/water mixing environment. Ceramics International. 2022;**48**: 4066-4073

[54] Huang Z, Rafiq M, Woldu AR, Tong QX, Astruc D, Hu L. Recent progress in electrocatalytic nitrogen reduction to ammonia (NRR). Coordination Chemical Reviews. 2023; **478**:214981

[55] Zhou H, Xiong B, Chen L, Shi J. Modulation strategies of Cu-based electrocatalysts for efficient nitrogen reduction. Journal of Materials Chemistry A. 2020;**8**:20286-20293

[56] Liu Y, Bai H, Zhang Q, Bai Y, Pang X, Wang F, et al. In-situ decoration of unsaturated Cu sites on Cu_2O photocathode for boosting nitrogen reduction reaction. Chemical Engineering Journal. 2021;**413**:127453

[57] Guo X, Yi W, Qu F, Lu L, et al. New insights into mechanisms on electrochemical N_2 reduction reaction driven by efficient zero-valence Cu nanoparticles. Journal of Power Sources. 2020;**448**:227417

[58] Yu J, Wang J, Ma Y, Zhou J, Wang Y, Lu P, et al. Recent progresses in electrochemical carbon dioxide reduction on copper-based catalysts toward multicarbon products. Advanced Functional Materials. 2021;**31**:2102151

[59] Zhang B, Zhang J, Hua M, Wan Q, Su Z, Tan X, et al. Highly electrocatalytic ethylene production from CO_2 on nanodefective Cu nanosheets. Journal of the American Chemical Society. 2020; **142**:13606-13613

[60] Choi C, Cai J, Lee C, Lee HM, Xu M, Huang Y. Intimate atomic Cu-Ag interfaces for high CO_2RR selectivity towards CH_4 at low over potential. Nano Research. 2021;**14**:3497-3501

[61] Zhang Q, Guan J. Applications of single-atom catalysts. Nano Research. 2021;**15**:38-70

[62] Zhang Y, Zhao J, Wang H, Xiao B, Zhang W, Zhao X, et al. Single-atom Cu anchored catalysts for photocatalytic renewable H_2 production with a quantum efficiency of 56. Nature Communications. 2022;**13**:58

[63] Cai Y, Fu J, Zhou Y, Chang YC, Min Q, Zhu JJ, et al. Insights on forming N,O-coordinated Cu single-atom catalysts for electrochemical reduction CO_2 to methane. Nature Communications. 2021;**12**:586

Chapter 6

Numerical Simulation of the Solar Cell Based on the Quaternary System $Cu(In,Ga)Se_2$

Daouda Oubda, Bawindsom Marcel Kébré, Soumaïla Ouédraogo, François Zougmoré and Zacharie Koalga

Abstract

Energy in one form or another has been and remains an essential vital element for humanity. It is positioned as the essential propellant for the accomplishment of any human activity. Its sources have diversified over time to meet the ever-increasing needs of industry and consumers. This chapter deals with the solar cell based on copper, indium, gallium, and diselenide (CIGS), which is a photovoltaic technology in constant evolution and demonstrates very interesting record yields. The aim is to show through the results obtained from the numerical simulation that the solar cells based on CIGS have the capacity in terms of energy conversion to meet part of the energy needs of the world population. In this chapter, after describing and modeling the Mo/Cu(In,Ga)Se_2/CdS/ZnO structure, we noted from the obtained results that Shockley-Read-Hall (SRH) recombination is the predominant mode of recombination in the CIGS solar cell, and these recombinations more affect the performance of the solar cell inside the space charge region (SCR).

Keywords: solar cell, CIGS absorber, CdS buffer layer, SCAPS-1D, operating temperature, series and shunts resistances, electrical parameters, SRH recombination

1. Introduction

Energy is at the center of all human activity; its transformation has been and still remains a great concern. It is increasingly used, especially in developed countries. Currently, more than 80% of the energy consumed in the world is of fossil origin [1–3]. In 2017, non-renewable energy sources constituted approximately 86% of global energy demand and nearly 48% of energy demand in Africa [4]. The rest of the energy (52%) used in Africa comes from essentially renewable energy in general and more particularly PV solar, wind, solar thermic but, above all hydraulic and biomass [4]. We see that fossil energy sources are limited in nature. In addition, the use of fossil fuels causes numerous social and environmental problems, including the degradation of the ozone layer, the main protector of the earth's atmosphere.

The move toward sustainable alternative energies is more relevant than ever [5]. The promotion of renewable energies, including photovoltaic, has become more than a necessity. This trend accelerated following the ratification of the Kyoto Protocol in 2002. Many countries then intensified research and development of alternative energies that do not emit carbon dioxide (CO_2) [5]. The Paris Agreement announced by the United Nations organization in 2015 on climate strengthens the Kyoto Protocol and aims to significantly reduce the risks and effects of climate change, and this involves limiting the rise in temperature at 1.5°C compared to pre-industrial levels. This will be made possible for low greenhouse gas emissions and will promote investment in the development of renewable energies. Solar photovoltaic (PV) energy is one of the most powerful alternatives for the future of large-scale electricity production. Solar cells based on $Cu(In,Ga)Se_2$ (copper, indium, gallium, and selenium) abbreviated CIGS from the second generation are among the most promising. They are developed with the aim of properly converting the maximum amount of light energy received into electrical energy, effectively reducing electrical and optical losses, and optimizing surface properties. It meets the current objectives in the field of photovoltaic: low cost, high conversion efficiency, mass production, and accessibility for all [6]. The development of a stable, high-efficiency solar cell remains the ultimate objective of researchers in the PV field. To improve their performance and increase their yields, we focus our study on the buffer layer.

In this chapter, the main objective is to contribute to a better understanding of the role of the cadmium sulfide (CdS) buffer layer and the recombination mechanisms in the operation of the thin-film solar cell based on CIGS.

The chapter is structured into several sections following the specific objectives of the study. The first section (2.1) addresses photovoltaic solar energy and photovoltaic sectors. In Section 2.2, we discuss the history of thin-film solar cells based on copper and indium diselenide. We further describe the structure of the solar cell, highlighting the role of each layer. We then discuss the operating principle of the solar cell and the effects of temperature on its operation.

Section 2.3 develops a numerical simulation methodology. We highlight the basic equations that govern the phenomenon of charge carrier transport in semiconductors and we discuss the process of solving these equations.

In Section 2.4, we emphasize the need for a numerical method and present simulation software. Section 2.5 describes the two solar cell characterization techniques used in this work, namely current-voltage density and quantum efficiency characterizations.

In the last section (2.6), we present the results of the numerical simulation with solar cell capacitance simulator in one-dimensional (SCAPS-1D) of the $Mo/Cu(In,Ga)Se_2/CdS/ZnO$ structure. We study the evolution of the charge carrier recombination rate across the thickness of the buffer layer. Likewise, we study the effects of the outside temperature on the performance of the solar cell. Finally, we determine the dominant recombination mechanism and its area of predominance.

2. Modeling and simulation of the Mo/CIGS/CdS/ZnO solar cell

2.1 Polycrystalline sector with thin layers of chalcopyrite

In the thin-film polycrystalline sector, three options stand out: cadmium telluride CdTe technology [5, 7, 8], copper indium diselenide technology $CuInSe_2$ (abbreviated CIS), and copper, zinc, tin, and selenium technology ($Cu_2ZnSnSe_4$) [9–12]. Copper and indium sulfide (CIS) is a ternary material of type I-III-VI, with a chalcopyrite

DOI: http://dx.doi.org/10.5772/intechopen.1003946

structure. It is increasingly promising for the mass production of thin-film modules, which is essentially explained by the interesting properties and the prospect of miniaturization of the solar device. The absorption coefficient of the (n)CdS-(p)$CuInSe_2$ heterojunction has approximately the same threshold as that of crystalline silicon (c-Si), but CIS is 100–1000 times more absorbent in the range of 1.1–2.6 eV than crystalline silicon. This allows an absorber thickness of 1–2 μm to absorb the maximum of photons from the solar spectrum [13, 14].

In the best current devices, Cu(In,Ga)Se_2 has a gap of the order of 1.2 eV obtained for [Ga]/[Ga] + [In]) =0.3 [15, 16]. In addition to the high conversion efficiencies that it promises, CIGS cells have good stability [1]. The interest in these thin layers comes essentially from the economical use of materials in relation to the physical properties and the simplicity of the technologies used for their production: easy and inexpensive development.

2.2 Solar cell based on the Cu(In,Ga)Se_2 quaternary system

2.2.1 Structure of the solar cell

We distinguish from bottom to top (**Figure 1**) [17–20]: substrate layer, rear contact layer, absorber layer, buffer layer, conductive transparent oxide layer, and front contact layer. To these elements is added a magnesium fluoride (MgF_2) anti-reflection layer.

2.2.1.1 Substrate

The soda glass substrate layer is the support of the device (**Figure 1**), and its thermal expansion coefficient is adapted to the growth of the CIGS layers. As early as 1993, the importance of contamination of the CIGS absorber by the sodium (Na) contained in the substrate was reported by Hedström et al. [16]. The sodium contained in the glass in the form of sodium oxide (Na_2O_2) plays a determining role both in the growth and performance of the cell [15, 21–23]. Sodium mainly improves the efficiency of the CIGS cell by increasing the open circuit voltage (V_{OC}) and the fill factor [22], other studies also reported an increase in the grain size.

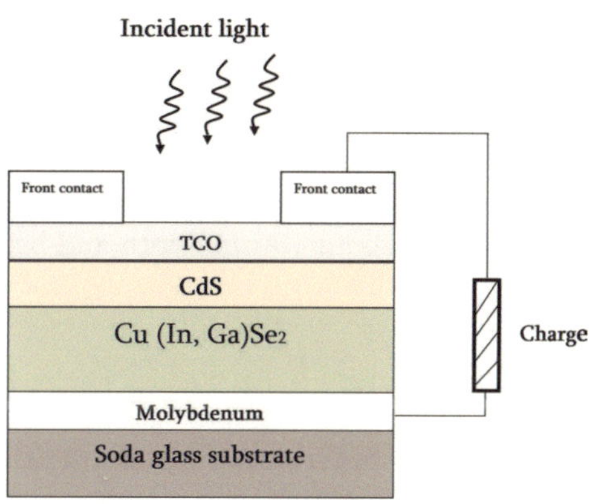

Figure 1.
Structure of a solar cell based on CIGS.

2.2.1.2 Back contact

The rear contact layer is made up of molybdenum (Mo) (**Figure 1**), it ensures the collection of positive charges that result from the conversion of incident photons into electron-hole pairs. The permeability of this layer allows the sodium atoms to diffuse and contribute to the better growth of the absorber at low and high temperatures [15, 22, 24–26]. Likewise, the selenium and gallium from the absorber diffuse into the molybdenum and form compounds such as molybdenum diselenide ($MoSe_2$) in very small quantities after annealing at 600°C [27–29]. The good ohmic and electronic behavior of the Mo/CIGS interface is due to the formation of the thin layer of $MoSe_2$ [23, 25, 28], which allows better adhesion of the molybdenum layer on the glass. The role of the $MoSe_2$ layer with a thickness of around 10 nm has been discussed within the scientific community [27, 29], and it is now proven that $MoSe_2$ is essential to facilitate the formation of a quasi-ohmic electrical contact at the interface between CIGS and molybdenum [27].

2.2.1.3 Absorber layer

The absorber layer, the most important of the device (**Figure 1**), is made up of indium gallium copper diselenide ($Cu(In,Ga)Se_2$). With p-type conductivity and a large absorption coefficient of around 10^5 cm^{-1} [1], it allows optimal conversion of incident photons into electron-hole pairs and ensures their separation. $CuInGaSe_2$ is the key material in CIGS solar cells, which is why its structural and optoelectronic properties have been and will be the subject of several studies. The performance of solar cells based on CIGS depends considerably on the structure of the material, the doping used, the insertion of gallium, or even the different interactions produced at the two interfaces of this layer, $CuInGaSe_2$/Mo and CdS/$CuInGaSe_2$.

2.2.1.4 Buffer layer

The buffer layer constituted by the cadmium sulphide (CdS) is n-type (**Figure 1**). It creates the heterogeneous p-n junction with the absorber layer [30]. Optimizing the performance of the CIGS-based solar cell necessarily requires understanding the mechanism of the formation of this p-n junction. CdS films have the advantage of having high optical transparency and low electrical resistivity, and the gap of 2.4 eV is among the largest [30]. The CdS layer passivates the defects on any free surface of the CIGS absorber and protects it during the deposition of the ZnO layer by cathode sputtering [25, 23]. The CdS layer provides a good quality interface with the CIGS layer and makes it possible to control the phenomenon of diffusion of defects toward this interface. It has been reported that the element cadmium present in cadmium sulfide diffuses into the absorber at about 10 nm depth and brings positive effects [31, 32].

2.2.1.5 Transparent conductive oxide layer

There are two main requirements for transparent conductive oxides (TCO) used in CIGS solar cells [16], sufficient transparency to allow enough light to reach the CIGS layer and conductivity high enough to be able to transport the photocurrent generated to an external circuit.

For our study, TCO is made up of zinc oxide (ZnO) (**Figure 1**). We distinguish n-ZnO doped TCO with a wide gap of 3.3 eV and high transparency. It mobilizes the maximum number of incident photons from the incident light toward the absorber. The presence of the intrinsic TCO i-ZnO in the solar cell based on Cu(In,Ga)Se_2 reinforces the protection of the absorber and also protects the buffer layer during the deposition of energetic ions from n-ZnO by sputtering cathodic. The i-ZnO layer has been shown to increase device performance due to increased open circuit voltage (V_{OC}) [16, 33].

2.2.1.6 Front contact

Finally, the front contact layer which is made up of a metal grid made of aluminum (Al) and nickel (Ni) ensures good collection of electrons (**Figure 1**). This metal grid is made up of a succession of layers (Ni/Al/Ni). The first layer of nickel prevents the oxidation of the aluminum contained in the TCO. The aluminum layer represents the front ohmic contact of the device, and the second nickel layer prevents oxidation of the aluminum, which could create direct contact with the ZnO window layer [22].

2.2.2 Operating principle

Regarding the CIGS-based solar cell, photons with energy Eph < 3.3 eV pass through the ZnO layer. Those with an energy between 2.4 and 3.3 eV are absorbed in the CdS buffer layer. In general, the largest majority of photons will reach the CIGS absorber to be absorbed in the Space Charge Region (SCR) [6, 17]. The photo-generated electron-hole pairs in the SCR are separated by an internal electric field, which then accelerates and propels the electrons toward the n-type zone and the holes toward the p-type zone [34]. The electrons are collected at the front contact (metallic grid) (**Figure 1**) and the holes at the back contact (Mo) (**Figure 1**). These carriers give rise to a generation photocurrent [35]. The electron-hole pairs that are collected at the metal grid (front contact), and at the ohmic contact (rear contact) of the cell are delivered into the load (**Figure 1**).

2.2.3 Influence of temperature

Temperature is a limiting parameter for solar photovoltaic devices. Cells exposed to solar radiation heat up. The very energetic photons transfer the energy equivalent to the gap of the semiconductor material, and the excess energy is dissipated in the form of heat through the device [3]. The increase in the temperature of the cell causes an expansion of the crystal lattice of the semiconductor material. The value of the gap decreases, which causes an increase in the saturation current [36]. Conversely, photons are more absorbed at long wavelengths, which leads to a slight increase in the short-circuit current [36]. We note that more free electrons are produced in the conduction band (BC) to the benefit of the slight reduction in the gap. The instability of the gap at high temperature can accelerate the recombination between the valence band (BV) and the (BC) [37], and the open-circuit voltage of the cell decreases. Operating at the lowest possible temperature allows for better conversion efficiency.

2.3 Modeling of the thin-film solar cell based on CIGS

The performance of a solar cell depends on its various electrical parameters, which are the short circuit current density (J_{SC}), the open circuit voltage (V_{OC}), the fill factor (FF), and its conversion efficiency photoelectric (η). Obtaining these different parameters of the solar cell requires the resolution of a certain number of equations. In this section, we will first highlight the essential equations governing the operation of thin-film solar cells. We then discuss solving these equations.

2.3.1 Transport phenomenon in a semiconductor

The differential equations governing the operation and characterization of a solar cell can be reduced to three (3) [35]:

- Poisson's equation:

$$\nabla.\varepsilon\nabla\Phi = -q\left(p - n + N_D^+ - N_A^-\right) \tag{1}$$

- the continuity equation for electrons:

$$\nabla.\vec{J}_n = q(R - G) + q\frac{\partial n}{\partial t} \tag{2}$$

- the continuity equation for holes:

$$\nabla.\vec{J}_p = -q(R - G) + q\frac{\partial p}{\partial t} \tag{3}$$

In these equations, ε is the dielectric constant, Φ the electrostatic potential, q the electrostatic charge; the terms n and p designate the concentrations of electrons and holes.

N_D^+ and N_A^- are the densities of ionized donors and ionized acceptors, respectively; the terms J_n and J_p are the current densities of the electrons respectively of the holes. R is the recombination rate, and G is the generation rate.

In a particular case of operation in a static regime, we have:

$$\frac{\partial n}{\partial t} = \frac{\partial p}{\partial t} = 0 \tag{4}$$

Then Eqs. (2) and (3) become:

$$\nabla.\vec{J}_n = q(R - G) \tag{5}$$

$$\nabla.\vec{J}_p = -q(R - G) \tag{6}$$

The recombination term in Eqs. (2) and (3) is a function of the charge carrier density. Therefore Eqs. (1)-(3) are nonlinear differential.

2.3.2 Generation term

In our work, generation results from optical excitation. The density of the photon flux denoted $\Phi(z)$, decreases exponentially with the depth z of the device. The back reflection shows that a fraction of the absorbed light reaches the back contact. If e_{CIGS} is the thickness of the absorber, the flux of the intensity of the reflected light has the expression [35]: $\Phi_{reflexie} = R_B \Phi_{e_{CIGS}}$, R_B is the reflection of the rear contact. The generation rate due to optical excitation is given by the following relationship:

$$G(z) = -\frac{d\Phi}{dz} = \alpha_x \Phi(z_0) \exp(-\alpha_x (z - z_0)) \tag{7}$$

In this equation, Φ is the photon flux density per unit area and time, and α_x is the absorption constant, the subscript x denotes a given layer: ZnO, CdS, and CIGS.

2.3.3 Recombination term

There are several types of recombination: surface, trap, Auger-type volume, radiative-type volume, and Schockley-Read-Hall recombination [38].

Schockley-Read-Hall recombination appears to be the predominant mode of recombination in the solar cell. It introduces a step in the transition between the conduction and valence bands in the form of an intermediate electronic state located in the forbidden band. The recombination rate R can be described by the following equation [35]:

$$R = \frac{(np - n_i^2)}{\tau_h(n + n_1) + \tau_e(p + p_1)} \tag{8}$$

This equation expresses the recombination in the quasi-neutral regions of a doped semiconductor. Under these conditions, the recombination rate of carriers in materials n and p is expressed respectively:

$$R_n = \frac{\Delta p}{\tau_h} \tag{9}$$

$$R_p = \frac{\Delta n}{\tau_e} \tag{10}$$

With Δn and Δp, the excess concentrations of electrons and holes.

2.3.4 Solving of equations

Numerical simulation is an effective tool used to understand and solve very complex problems related to the structure of solar cells. It makes it possible to effectively model and improve the conversion efficiency of CIGS-based solar devices.

The unknowns of Eqs. (1)-(3) are the following state variables: the electrostatic potential Φ, the quasi-Fermi level energies for electrons, and holes E_{fn} and E_{fp}. These three state variables are sufficient to describe the operation of the solar cell. The occupation of the conduction band by electrons and of the valence band by holes can be described by the Fermi-Dirac statistic:

$$f(E) = \frac{1}{1 + \exp\left(\frac{E - E_f}{kT}\right)} \tag{11}$$

This function describes the probability of occupation of the electron in the conduction band and the expression $1 - f(E)$ that of the hole in the valence band. With k the Boltzmann constant, T the temperature, E the energy of the electron for a given level, and E_f the energy for the Fermi level. The occupation of the Fermi level is given by the expressions for the densities of the donors n and that of the acceptors p [39]:

$$n = N_C \exp\left(-\frac{E_C - E_f}{kT}\right) \tag{12}$$

$$p = N_V \exp\left(-\frac{E_f - E_V}{kT}\right) \tag{13}$$

N_C and N_V are the effective densities of electrons in the conduction band and the effective densities of holes in the valence band and have the following expressions respectively:

$$N_C = 2\left(\frac{2\prod m_e^* kT}{h^2}\right)^{3/2} \tag{14}$$

$$N_V = 2\left(\frac{2\prod m_h^* kT}{h^2}\right)^{3/2} \tag{15}$$

In this equation, h is Planck's constant, m_e^* and m_h^* are the effective masses of the electron and the hole. In equilibrium, the product of n and p is constant and depends only on the temperature, the effective masses, and the gap of the semiconductor:

$$np = n_i^2 = N_C N_V \exp\left(-\frac{E_g}{kT}\right) \tag{16}$$

In non-equilibrium situations, such as under illumination or during fault injection, there is no uniform Fermi level. In the steady state, the energies of the quasi-Fermi levels E_{fn} and E_{fp} can be introduced and the product np depends on the voltage:

$$np = n_i^2 \exp(qV/kT) \tag{17}$$

In the absence of a magnetic field and temperature gradient, the transport of a charge carrier in semiconductors occurs by diffusion and is expressed from the following equations relating to the current density of electrons and holes [35, 39]:

$$J_n = q(\mu_n n\varepsilon + D_n \nabla n) \tag{18}$$

$$J_p = q\left(\mu_p p\varepsilon - D_p \nabla p\right) \tag{19}$$

With μ_n andμ_p, the mobility of electrons and holes, respectively; the terms D_n and D_p, diffusion constants of electrons and holes, respectively.

The introduction of the Fermi potential Φ_{fn} and Φ_{fp}, respectively for electrons and holes as a function of the energy of the quasi Fermi level E_{fn} and E_{fp}, makes it possible to simplify Eqs. (18) and (19):

$$\Phi_{fn} = -E_{fn}/q \tag{20}$$

$$\Phi_{fp} = E_{fp}/q \tag{21}$$

$$J_n = -q\mu_n n \nabla \Phi_{fn} \tag{22}$$

$$J_p = -q\mu_p p \nabla \Phi_{fp} \tag{23}$$

Expressions (20) and (21), applied to Eqs. (1)-(3), describe the equation for the transport of charge carriers in the structure.

Eqs. (1)-(3) are subject to the conditions on the potential Φ and the current density J at the contacts. In a one-dimensional case, we have the following expressions [40]:

$$\Phi(0) = \Phi_0 - V \tag{24}$$

$$\Phi(L) = 0 \tag{25}$$

$$J_p(0) = -qS_{p0}(p_0(0) - p(0)) \tag{26}$$

$$J_p(L) = qS_{pL}(p(L) - p_0(L)) \tag{27}$$

$$J_n(0) = qS_{n0}(n(0) - n_0(0)) \tag{28}$$

$$J_n(L) = -qS_{nL}(n_0(L) - n(L)) \tag{29}$$

$p_0(0)$ and $p_0(L)$ are the populations (density of holes) of the valence band respectively at the position $X = 0$ and the thermodynamic equilibrium position $X = L$. $n_0(0)$ and $n_0(L)$, the electron density of the conduction band at $X = 0$ and at thermodynamic equilibrium $X = L$. The quantities $p(0)$ and $p(\mathrm{L})$ correspond to the population of holes encountered at $X = 0$ and $X = L$. The quantities $n(0)$ and $n(p)$ correspond to the population of electrons found in the region delimited by $X = 0$ and $X = L$. The terms S_{P0}, S_{pl}, S_{n0}, and S_{nL} are the effective recombination speeds on the surface of holes and electrons at the respective positions $X = 0$ and $X = L$ [40].

2.4 Problematic of solving equations in thin layers

2.4.1 Need for a numerical method

Eqs. (1)-(3) are highly non-linear partial differential equations (PDE). The fact of having several nested layers, each with different input parameters, complicates the equations and prevents analytical solutions. These input parameters are of three types:

- the parameters that act on the entire device such as the temperature, the speed of thermal agitation, and the reflection coefficient;
- the parameters that apply to a particular region of the device such as the properties of the materials, the mobility of electrons and holes, and the width of the gap;
- the parameters that apply to the light spectrum such as light, wavelength, and absorption coefficient.

A complete solution to Eqs. (1)-(3) must correctly describe the possible discontinuities in the energy bands at the layer interface. Likewise, the effect of deep energy states (defects) in the majority of layers must be fully addressed and an explanation regarding recombination in the space charge region must be provided. In addition, the solution to Eqs. (1)-(3) must allow the determination of the state variables, $\varepsilon(x)$, $n(x)$, and $p(x)$, which completely define the system at each point x. Given all of the above, an analytical solution to this problem is then excluded. Numerical methods appear essential for solving these equations. A typical methodology includes the following approach:

- discretization (mesh) of the device;
- discretization of the Poisson equation and the continuity equations for electrons and holes;
- application of contact boundary conditions and initial conditions;
- solution of the resulting matrix equation by iteration (Newton-Raphson method, Picard).

In short, a simulation program for thin-film solar cells must take into account the presence of several layers, and the recombination phenomena in volume and at the interface of the layers must be considered. It must also correctly deal with the recombination problem and the generation-recombination centers in the deep states of the volume of the layers. It must also offer the simulated opto-electrical characteristics (J-V, QE-λ, C-V, and C-f) and allow a comparison with the experimental measurements carried out.

2.4.2 Simulation software

Among the solar cell simulation software, we have AMPS, PC, SCAPS, and AFORS-HET at our disposal. Software is used directly or indirectly to evaluate the performance of the cell in one dimension [41]. All software in its diversity uses the equations described above. In the following lines, we will describe AMPS and SCAPS. The latter two are used specifically for chalcopyrite thin-film solar cells.

2.4.2.1 Analysis of microelectronic and photonic structures-1D

Analysis of microelectronic and photonic structures-1D (AMPS-1D) is the work of the team of Professor Stephen J. Fonash, from Pennsylvania State University with the support of the Electric Power Research Institute [1]. It makes it possible to obtain the response of any optoelectronic device to a given excitation: light, bias voltage, or temperature [42]. Among these devices, we distinguish:

- homo-junctions and heterojunctions p-n, p-i-n, n-i-n, and p-i-p;
- microelectronic structures;
- detector structures, and solar cell structures.

In the numerical simulation, AMPS-1D takes into account several layers in the case of heterojunctions. The simulation of a solar cell with AMPS-1D gives the current density-voltage characteristics under illumination and darkness as well as the charge collection efficiency as a function of voltage. Furthermore, important information such as free and trapped electric field distributions, recombination profiles, and densities of each current carrier as a function of position are extracted from the simulation.

2.4.2.2 Solar cell capacitance simulator

Solar cell capacitance simulator (SCAPS-1D) is the work of Marc Burgelman's team from the Department of Electronics and Information Systems (ELIS) at the Ghent University in Belgium. SCAPS-1D, one-dimensional numerical simulation software, is developed to obtain the electrical characteristics of heterojunction and thin-film solar cells. Its development is inspired by work on solar cells based on CdTe and Cu(In,Ga) Se_2 [19, 43]. The simulated and measured results were in very good agreement with the results in the literature [20, 41]. With SCAPS, it is possible to simulate structures consisting of seven intermediate layers, front and back contacts with different doping profiles, and energy distributions of given donor or acceptor levels. All physical and electronic properties can be viewed and edited in a separate window. SCAPS-1D, through numerical simulation, studies the most sensitive parameters of the solar cell, namely J(V), C(V), C(f), and QE(λ) under illumination and in the dark [43]. This feature makes SCAPS a very attractive program that catches our attention.

2.5 Characterization of a thin-film solar cell based on CIGS with SCAPS-1D

Characterization of CIGS-based solar devices is necessary to determine the sources of losses and suggest ways to minimize them. We investigate two characterization techniques in this manual: J-V and QE-λ.

2.5.1 Characterization of current density-voltage J-V

The current density-voltage characteristic of the solar cell is described by a standard diode equation. The diode current density is limited by Shokley-Read-Hall recombinations across states of the CIGS space charge zone and leads in the simplified case to the following equation [4]:

$$J = J_0 \exp[qV/AkT] - J_{ph} \tag{30}$$

J_0 is the diode saturation current density, A is the quality factor, J_{ph} is the photogenerated current density ($J_{ph} = J_{SC}$), k is the Boltzmann constant, and T is the temperature. The simplest and most effective way to describe the operation of a solar cell is to use its current density-voltage characteristics. There are two modes of current density-voltage characterization (**Figure 2**):

- Under darkness, the J-V curve follows the exponential diode law. SCAPS makes it possible to monitor the effects of series and shunt resistances on the performance of the solar cell.

- Under illumination, the J-V curve is shifted downward by an amount designated by J_{ph} called photogenerated current density, which in the ideal case is independent of the open-circuit voltage (V_{OC}) and the short current density circuit (J_{SC}). SCAPS allows us to directly access the main electrical parameters which are J_{SC}, V_{OC}, FF, and η.

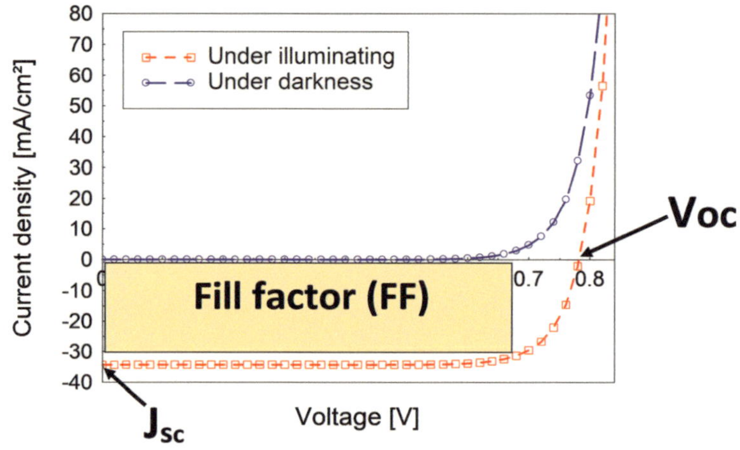

Figure 2.
Characteristics (J-V) under darkness and illumination [4].

2.5.2 Quantum efficiency.

Quantum efficiency (QE) or quantum yield is defined as the ratio of charge carriers collected to photons incident on a surface at each wavelength. In the ideal case, each photon creates an electron-hole pair that contributes to the photocurrent. For an ideal cell, the highest energy photons ($h\upsilon > Eg$) give an efficiency of 100% and 0% for the lower energy photons ($h\upsilon < Eg$). This is not the case for a real cell because of reflection losses and losses due to incomplete collection of charge carriers (**Figure 3**). Photons whose energy is lower than the band gap therefore do not contribute to the current density.

Losses at short wavelengths are the result of absorption in the CdS layer ($\lambda < 520$ nm) and in the ZnO layer ($\lambda < 375$ nm) because most of the carriers generated

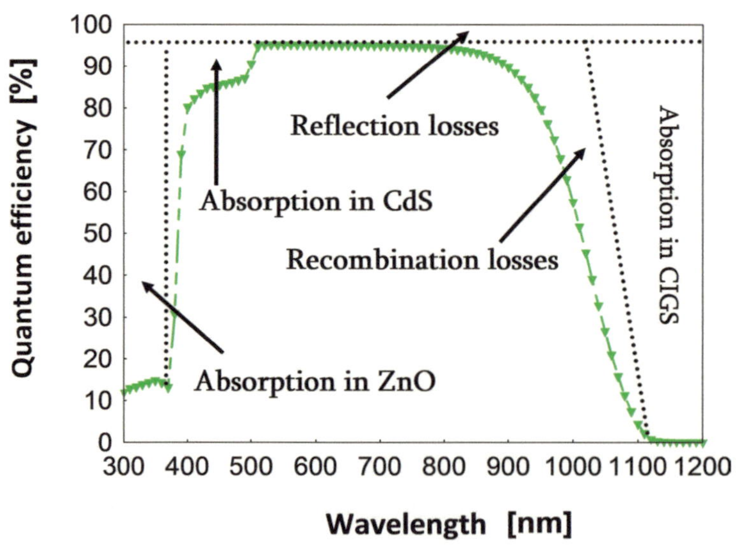

Figure 3.
Quantum efficiency with current losses.

in these layers are generally not collected [1]. The spectral response is the ratio between the density of the collected current and the density of the incident photons, for each wavelength of light radiation. It depends much more on the optical properties of the materials than on the spectral distribution of the light received; it makes it possible to evaluate the quantum efficiency of the cell as a function of the wavelength of the incident light. Its optimization requires improvement in the front and even rear surfaces.

The internal spectral response (IQE) and the external spectral response (EQE), are strongly linked to surface and volume recombination, the diffusion length of the carriers, and the thickness of the region concerned [16]. External quantum efficiency takes into account optical loss effects such as unabsorbed light and reflected light. It is given by the equation:

$$RQE(\lambda) = \frac{J_{ph}(\lambda)}{q\Phi_0(\lambda)} \tag{31}$$

The ratio between the number of electron-hole pairs collected and the number of absorbed photons provided is the internal spectral response [1]. It makes it possible to evaluate the quality of the different regions of the solar cell and is given by Eq. (32) [36, 44].

$$RQI(\lambda) = \frac{RQE(\lambda)}{(1 - R(\lambda))} = \frac{1}{(1 - R(\lambda))} \frac{J_{ph}(\lambda)}{q\Phi_0(\lambda)} \tag{32}$$

These two forms of spectral response are strongly linked to surface and volume recombination, the carrier diffusion length, and the thickness of the layers of the solar cell [16].

2.6 Results of the numerical simulation with SCAPS-1D

2.6.1 Study of the CdS buffer layer

In this part, we vary the thickness of the buffer layer and we obtain the curves relating to the electrical and optical characteristics respectively, J-V and QE-λ.

2.6.1.1 Influence of the thickness of the CdS on the electrical characteristics

At T = 300 K, we obtain **Figure 4** and **Table 1** which represent the influence of the thickness of the buffer layer on the J-V characteristic.

From the results in **Table 1** and **Figure 4**, we see a decrease in the values of all electrical parameters with the increase in the thickness of the CdS buffer layer [6]. However, we note a weak influence of the thickness of the CdS layer on the V_{OC}. In addition, the results show that the values of all the electrical parameters are greater for the small thickness of the CdS.

These results can be explained, on the one hand, by a reduction in the quantity of incident photons arriving at the absorber. This decrease is due to the absorption of incident photons in the CdS layer, which becomes important with the increase in the thickness of the CdS [6]. On the other hand, the value of the series resistance increases in the solar cell, which increases with the increase in the thickness of the CdS.

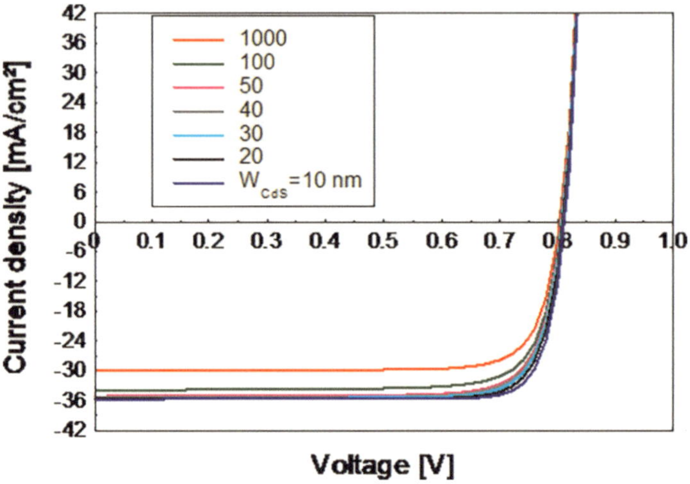

Figure 4.
J-V characteristic as a function of the thickness of the CdS [6].

W_{CdS} (nm)	0	10	20	30	40	50	100	1000
V_{OC} (V)	0.806	0.796	0.787	0.782	0.779	0.778	0.776	0.774
J_{SC} (mA/cm^{-2})	35.74	35.28	34.77	34.30	33.88	33.51	32.14	29.83
FF (%)	85.34	83.69	81.71	79.80	78.37	77.54	76.46	76.28
η (%)	24.60	23.51	22.38	21.42	20.70	20.22	19.27	18.27

Table 1.
Values of electrical parameters depending on the thickness of the CdS at 300 K [6].

2.6.1.2 Influence of the thickness of the CdS on the quantum efficiency

Figure 5 shows that for small thicknesses ($10 \leq W_{CdS} \leq 50$ nm), the range of wavelengths absorbed in the CIGS layer goes from 380 nm to 1100 nm. When the thickness of the CdS is greater than 50 nm, the absorption in the buffer layer becomes very important. For a sufficiently large thickness (1000 nm) of the CdS, we see from **Figure 5** that the length limiting absorption wave of the buffer layer is approximately 500 nm. As a consequence, all incident photons with a wavelength less than or equal to 500 nm are absorbed in this layer (CdS) [6].

Unfortunately, the majority of electron-hole pairs photogenerated in the CdS layer recombine and are not collected at the contact levels. This explains the low values of J_{SC} and η observed in **Table 1**. We emphasize that the significant reduction in conversion efficiency (**Table 1**) is a direct consequence of the losses linked to J_{SC} and V_{OC} [6]. It appears from the study carried out on the buffer layer that the recombination inside the CdS layer and the interface of the CdS/CIGS layers become more and more important when the thickness of the CdS increases [19]. A thin layer of 10–40 nm of CdS is important for obtaining better performance and good stability [6].

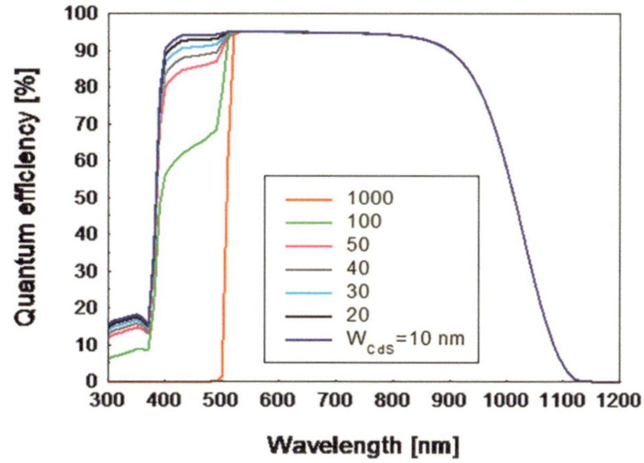

Figure 5.
Spectral response as a function of the thickness of the CdS [6].

2.6.2 Influence of temperature on the performance of the CIGS solar cell

2.6.2.1 J-V and quantum efficiency characteristics

Temperature is a limiting factor for the proper functioning of the solar cell. The present work effectively demonstrates that the J-V characteristic is affected by the temperature, in particular the open-circuit voltage of the device. We observed from the J-V characteristic (**Figure 6a**) [6], a significant decrease in V_{OC} with the increase in temperature. From the results in **Table 2**, in addition to V_{OC}, the values of FF and η decrease with increasing temperature [6], on the other hand, J_{SC} increases slightly.

2.6.2.2 Electricals parameters

We assume that a reduction in the quantity of photons absorbed at the absorber or a significant recombination of electron-hole pairs inside the SCR may be responsible

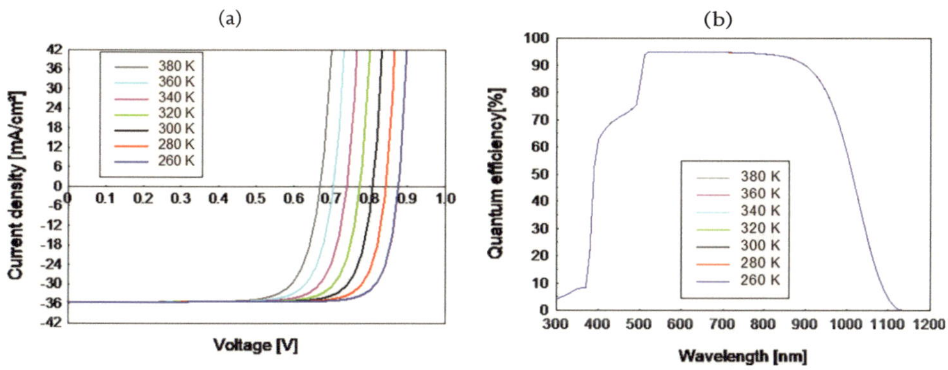

Figure 6.
(a) Current density-voltage characteristics and (b) quantum efficiency for different temperatures [6].

T (K)	260	280	300	320	340	360	380
V_{OC} (V)	0.8491	0.8159	0.7824	0.7482	0.7139	0.6794	0.6441
J_{SC} (mA/cm^{-2})	34.300	34.301	34.301	34.303	34.304	34.307	34.310
FF (%)	82.004	80.923	79.805	78.660	77.428	76.096	74.701
η (%)	23.890	22.652	21.419	20.189	18.96	17.732	16.504

Table 2.
Evolution of electrical parameters as a function of temperature [6].

for these results. However, the quantum efficiency of the solar cell appears not to be affected by the temperature variation as shown in **Figure 6b**. The quantity of photons absorbed in the absorber is therefore not reduced [6]. In reality, when the temperature increases, the mobility of the charge carriers decreases according to Eq. (3) and leads to significant recombination of the charge carriers. In addition, at high temperatures the gap of semiconductor materials decreases and leads to low performance of the solar cell. The slight increase in J_{SC} values is due precisely to this reduction in the gap of the absorber with temperature (**Table 2**).

$$\mu_{n,p} = D\frac{q}{kT} \tag{33}$$

The analysis of these results shows that the losses in the performance of our solar cell model as a function of temperature are linked to the recombination of electron-hole pairs inside the SCR. The performance losses at the maximum power point are approximately: 0.35–0.5%/°C for a solar cell based on crystalline silicon [36, 45]; 0.27–0.32%/°C for a CIGS-based solar cell [37]. For our model, losses as a function of temperature are estimated at 0.1%/°C [34].

2.6.3 Recombination in CIGS solar cells

In this part of our chapter, our objective is to determine the type of recombination mechanism (Auger, radiative, or SRH) dominant in our solar cell model. In addition, we seek to understand in which part of the cell (CdS layer, SCR, CIGS/Mo, or CdS/CIGS interfaces) this type of recombination mechanism most affects the performance of the solar cell.

2.6.3.1 Dominant recombination mechanism

Charge carrier recombination plays a major role in CIGS-based solar cells. Controlling the different recombination mechanisms is very important for better performance of solar cells. The recombination of a charge carrier depends on its lifetime (τ), its mobility (μ), its diffusion length (L), and its recombination speed (υ_{th}). There are radiative, Auger, and Shockley Read Hall (SRH) type recombination mechanisms. The simulation with SCAPS-1D allows us to obtain **Figure 7**, which highlights the effects of the different recombination mechanisms on the J-V characteristic. From **Figure 7**, we see a significant loss of J_{SC} by SRH-type recombination. We deduce from these results that SRH type recombinations are dominant within the present solar cell model. Based on the results of **Figure 7** we affirm that the different recombinations encountered in the previous sections are of the SRH type.

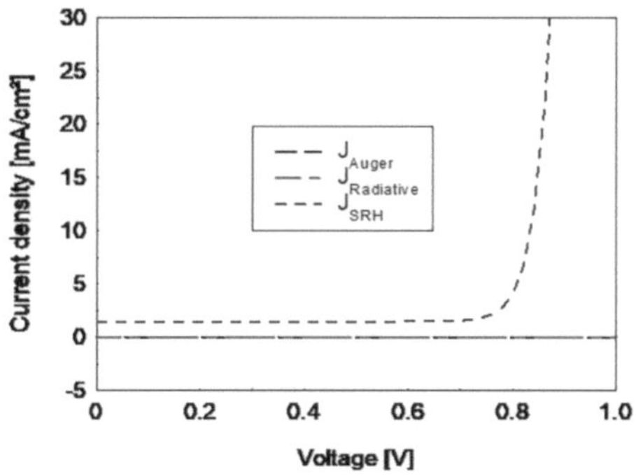

Figure 7.
Recombination current density.

2.6.3.2 Activation energy

The activation energy, denoted Ea, is a phenomenological parameter used to locate the place of the predominance of SRH-type recombination mechanisms in the CIGS solar cell. In the literature, several methods exist for determining the recombination activation energy [16, 17, 46, 47]:

The method that inspired us consists of plotting the V_{OC} curve as a function of temperature; we graphically obtain the activation energy by extrapolating the curve up to the temperature of 0 K at the origin of the benchmark (**Figure 8**). The V_{OC} value read corresponds to the value of the activation energy to a factor of 1/q (V_{OC} = Ea/q) (**Figure 8**). We make the following hypotheses:

1. If the activation energy is lower than the absorber gap (Ea < Eg), the interface recombination mechanisms (CdS/CIGS, CIGS/Mo) are dominant;

2. If the activation energy is substantially equal to or greater than the gap of the absorber, the recombination mechanisms inside the SCR dominate.

From **Figure 8** obtained from the numerical simulation, we see that the activation energy of the model studied has a value of approximately 1.3 eV. This value is greater

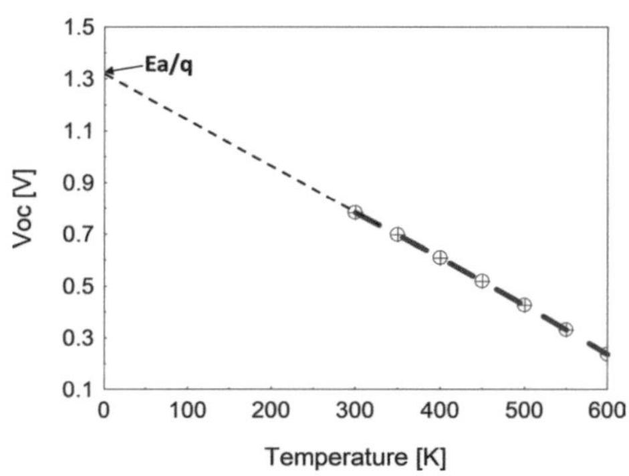

Figure 8.
Method for determining the activation energy.

than the gap value of the absorber studied (E.g. = 1.2 eV). In conclusion, the present model is affected more by SRH type recombinations in volume, more precisely inside the SCR.

3. Conclusion

In this manuscript, we proposed a theoretical framework for characterizing the CIGS-based solar cell based on the CdS buffer layer. We have set ourselves the objective of clearly explaining the different recombination phenomena that affect the performance of CIGS-based PV solar cells.

We described the polycrystalline sector with thin chalcopyrite layers and we presented the role of the different layers constituting the CIGS-based structure. Then, we discussed the operating principle as well as the influence of temperature on the Mo/CIGS/CdS/ZnO structure.

In the following sections, we modeled the CIGS-based solar cell from the basic equations that govern the transport phenomenon of charge carriers in a semiconductor. We also explained the terms of generation and recombination and underlined the need for a numerical method for solving the equations. We then gave an overview of the basic concepts of numerical simulation and described the SCAPS-1D and AMPS-1D simulation software. Finally, we presented the J-V and QE characterization techniques used for the numerical simulation.

We finally presented the results on the Mo/CIGS/CdS/ZnO structure in numerical simulation. The results show that the presence of a large thickness of the buffer layer reduces the quantity of photons arriving at the absorber. It also leads to recombinations inside the CdS layer of photogenerated charge carriers. We found that the electrical parameters decrease with increasing temperature, with the exception of J_{SC}, which observes a slight increase. In addition, the increase in temperature leads to recombination inside the SCR and greatly limits the performance of solar cells. Overall, we were able to show that SRH-type recombination is the predominant mode of recombination in CIGS-based solar cells. Our results further show that recombinations inside the SCR affect the performance of the solar cell.

It appears from this study that for good performance of our model, a thickness of 30 nm of the CdS layer is necessary. We retain the range of 300–360 K for the proper functioning of the chalcopyrite cell based on the $Cu(In,Ga)Se_2$ quaternary system.

Acknowledgements

The authors acknowledge the use of SCAPS-1D program developed by Marc Burgelman and colleagues at the Ghent University in all the simulations reported in this paper.

Conflict of interest

The authors declare no conflict of interest.

Author details

Daouda Oubda*, Bawindsom Marcel Kébré, Soumaïla Ouédraogo, François Zougmoré and Zacharie Koalga
M.E.LA. (Materials and Environment Laboratory), TRU-EAS (Training and Research Unit in Exact and Applied Sciences), Doctoral School of Sciences and Technologies Joseph KI-ZERBO University, Ouagadougou, Burkina Faso

*Address all correspondence to: daoudaoubd@gmail.com; daouda.oubda@ujkz.bf

References

[1] Kanevce A. Anticipated performance of Cu(In,Ga)Se_2 solar cells in the thin film limit [Ph.D. thesis]. Colorado State University; 2007

[2] Dupuis J. Development and characterization of SiOxNy:H and SiNx: H layers produced by the PECVD method: Application to the rear face of silicon photovoltaic cells [Ph.D. thesis]. INSA Lyon: Lyon Institute of Nanotechnologies; 2009

[3] Mostefa KS. Study and simulation of thin film photovoltaic cells based on CIS and CIGS [M.Ph. thesis]. Abou-Bakr Blekaid Tlemcen University; 2012

[4] Oubda D, Ouédraogo S, Diasso A, Kébré MB, Zougmoré F, Koalga Z, et al. Renewable energies in Africa, context and socio-economic challenge. Open Journal of Applied Sciences. 2022;**12**:1856-1872. DOI: 10.4236/ojapps.2022.1211128

[5] Yan X. Fabrication and characterization of CuInGaSe_2 films by cathode sputtering Study of defects by charge-based deep trap spectroscopy [Ph.D. thesis]. University of Nantes; 2014

[6] Oubda D, Kébré BM, Zougmoré F, Njomo D, Ouattara F. Numerical simulation of Cu(In,Ga)Se_2 solar cells performances. Journal of Energy and Power Engineering. 2015;**9**:1047-1055

[7] Lechat I. Experimental study and development of a prototype for characterizing a dye-based solar cell [Tech. rept]. University of Reunion; 2007

[8] Christophe M. New efficiency records for photovoltaic solar cells [Tech. rept]. 2016. Available from: https://www.notre-planete.info

[9] Munoz-Garci MA, Marin O, Alonso-Garci MC, Chenlo F. Characterization of thin film PV modules under standard test conditions: Results of indoor and outdoor measurements and the effects of sunlight exposure. Solar Energy. 2012; **86**:30493056

[10] Wätjen J. T. An electron microscopy study of Cu(In,Ga)Se_2 and Cu_2ZnSn(S, Se)$_4$ solar cells [Ph.D. thesis]. Uppsala University; 2013

[11] Mauguit Q. Photovoltaic cells on solar panels [Tech. rept]. Futura-Sciences; 2013

[12] Jie Z, Bo L, Shuying C, Weibo Z. Effects of sulfurization temperature on properties of CZTS films by vacuum evaporation and sulfurization method. International Journal of Photoenergy. 2013;**2013**(6):986076. DOI: 10.1155/2013/986076

[13] Ricaud A. Photovoltaic modules [Tech. rept.]. CYTHELIA; 2009

[14] Maoucha A. Study and parametric identification of an organic photovoltaic cell [M.Phil thesis]. University of Batna; 2011

[15] Kessler F, Dominik R. Technological aspects of flexible CIGS solar cells and modules. Solar Energy. 2004;**77**: 685695

[16] Ouédraogo S. Numerical modeling of a thin-film solar cell based on CIGS [thesis]. Ouagadougou: University of Ouagadougou; 2016

[17] Oubda D. Characterization of a thin film solar cell based on CIGS according to the nature of the buffer layer [thesis]. Ouagadougou: University of Ouagadougou; 2018

[18] Zongo A, Oubda D, Ouédraogo S, Kébré MB, Diasso A, Sankara I, et al. Optimization of Mo/Cu(In, Ga)Se_2/CdS/ZnO hetero-junction solar cell performance by numerical simulation with SCAPS-1D. Journal of Materials Science and Engineering B. 2021;**11**:156-167. DOI: 10.17265/2161-6221/2021.10-12.00

[19] Oubda D, Kébré MB, Ouédraogo S, Diasso A, Zougmoré F, Koalga Z, Ouattara F. High performance for Cu(In, Ga)Se_2 quaternary system-based solar cells with alternative buffer layers. Advances in Materials Physics and Chemistry. 2022;**12**:207-219. https://doi.org/10.4236/ampc.2022.129015

[20] Traoré B, Ouédraogo S, Oubda D, Kébré BM, Zongo A, Sankara I, Zougmoré F. Enhancing strategy CIGS solar cell performance through a new ZnSe buffer layer. Physics and Materials Chemisty. 2023;**9**:1:1-7. DOI: 10.12691/pmc-9-1-1

[21] Pudov OA. Impact of secondary barriers on $CuIn_{1-x}Ga_xSe_2$ solar-cell operation [thesis]. Colorado State University; 2005

[22] Painchaud T. Growth mechanisms of coevaporated Cu(In,Ga)Se2 thin films: Towards rapid and low-temperature syntheses [thesis]. University of Nantes; 2010

[23] Duchatelet A. Synthesis of thin layers of Cu(In,Ga)Se_2 for solar cells by electrodeposition of mixed copper-indium-gallium oxides [thesis]. Lille1 University; 2012

[24] Benaïcha M. Elaboration of ternary thin layers by electrochemical voice. Physical-chemical and electrical characterization [thesis]. Ferhat Abbas University of Sétif; 2010

[25] Buffière M. Synthesis and characterization of thin layers of Zn(O, S) for application in solar cells based on Cu(In,Ga)Se_2 [thesis]. University of Nantes; 2011

[26] Froger V. Thin films of zinc chalcogenides deposited by spray-CVD assisted by infrared radiation for photovoltaic applications [thesis]. National School of Arts and Crafts; 2012

[27] Abou-Ras D, Kostorz G, Bremaud D, Kelin M, Kurdesau FV, Tiwari AN, et al. Formation and characterisation of $MoSe_2$ for Cu(In,Ga)Se_2 based solar cells. Thin Solid Films. 2005;**480-481**(433):438

[28] Bollero A, Kaupmees L, Raadik T, Grossberg M, Fernández S. Thermal stability of sputtered Mo/polyimide films and formation of $MoSe_2$ and MoS_2 layers for application in flexible Cu(In, Ga)$(Se,S)_2$ based solar cells. Thin Solid Films. 2011;**520**(12):4163-4168. DOI: 10.1016/j.tsf.2011.04.099

[29] Wei-Ting L, Shih-Hao C, Shao-Ze T, Jhih-Jian H, Sheng-Hui C, Ruei-Fu S, et al. Manipulation of $MoSe_2$ films on CuIn(Ga)Se_2 solar cells during rapid thermal process. International Journal of Photoenergy. 2014;**2014**:253285. DOI: 10.1155/2014/253285

[30] Keriya M. Tem and Cathodoluminescence investigations of CdS grown by different techniques [thesis]. Durham Unuversity; 2005

[31] Nakada T, Kunioka A. Direct evidence of Cd diffusion into Cu(In,Ga)Se_2 thin films during chemical-bath deposition process of CdS films. Applied Physics Letters. 1999;**74**:2444-24466

[32] Hariskos D, Fuchs B, Menner R, Naghavi N, Hubert C, Lincot D, et al. The Zn(S,O,OH)/ZnMgO buffer in thin-

film Cu(In,Ga)(Se,S)$_2$-based soletar cells part II: Magnetron sputtering of the ZnMgO buffer layer for in-line co-evaporated Cu(In,Ga)Se$_2$ solar cells. Progress in Photovoltaics: Research and Applications. 2009;**17**:479-488

[33] Rau U, Marion S. Electronic properties of ZnO/CdS/Cu(In,Ga)Se$_2$ solar cells aspects of heterojunction formation. Thin Solid Films. 2001;**387**: 141-146

[34] Oubda D, Kebre MB, Ouédraogo S, Zougmoré F, Ouattara F, Koalga Z. Numerical characterization of Cu(In,Ga) Se$_2$ solar cells using capacitance-voltage and capacitance-frequency characteristics. International Journal of Progressive Sciences and Technologies (IJPSAT). 2018;**6**(2):262-267

[35] Gloeckler M. Device physics of Cu (In,Ga)Se$_2$ thin-film solar cells [thesis]. Colorado State University; 2005

[36] Veschetti Y. Modeling, characterization and production of new photovoltaic structures on thin silicon substrate [thesis]. Louis Pasteur University Strasbourg I; 2005

[37] Puvaneswaran C, Mohammad IH, Nowshad A. Performance analysis of copper- indium-gallium-diselenide (CIGS) solar cells with various buffer layers by SCAPS. Current Applied Physics. 2010:S387S391

[38] Mathieu H. MASSON. Physics of Semiconductors of Electronic Components. 6th edition. 1987. pp. 88-90

[39] Shockley W, Sparks M, Teal G. p-n junction transistors. Physics Review. 1951;**83**:151-162

[40] Diouf D. Silicon photovoltaic cells with heterojunctions and an interdigitated structure on the rear face [thesis]. University of Paris Sud-11; 2010

[41] Burgelman M, Nollet P, Degrave S. Modelling polycrystalline semiconductor solar cells. Thin Solid Film. 2000; **361-362**:527-532

[42] Jiyon S, Sheng S, Li C, Huang O, Crisalle T, Anderson J. Device modeling and simulation of the performance of Cu (In1-x,Gax)Se2 solar cells. Solid-State Electronics. 2004;**48**:7379

[43] Niemegeers A, Burgelman M, Herberholz R, Rau U, Hariskos D, Schock H. Model for electronic transport in Cu(In,Ga)Se$_2$ solar cells. Applied Physics Letters. 1998;**6**:407-421

[44] Pettersson J. Modelling band gap gradiens and Cd-free buffer layers in Cu (In,Ga)Se$_2$ solar cells [thesis]. Uppsalat University; 2012

[45] Ricaud A. Photovoltaic modules and systems [Tech. rept.]. CYTHELIA; 2008

[46] Rusu M, Eisele W, Würz R, Ennaoui A, Lux-Steiner MC, Niesen TP, et al. Current transport in ZnO/ZnS/Cu (In,Ga)(S,Se)$_2$ solar cell. Journal of Physics and Chemistry of Solids. 2003; **64**:2037-2040

[47] Hultqvist A. Cadmium free buffer layer and the inflence of their material properties on the performance of Cu(In, Ga)Se$_2$ solar cells [thesis]. Uppsala Universitet; 2010

Section 4

Other Copper Applications

Chapter 7

Copper Application and Copper Nanoparticles in Chemistry

Iman Mohammadi Dehcheshmeh, Ahmad Poursattar Marjani, Fatemeh Sadegh and Mohammad Ebrahim Soltani

Abstract

Copper metal is a natural element found in soil, water, and rocks. This metal is one of those functional metals that have significantly improved the quality of human life. In the agricultural industry, copper plays an essential role as a primary nutrient required for the optimal growth of living tissues in plants and other organisms. Additionally, it is used to control fungal diseases; copper sulfate, one of the most widely used derivatives of copper metal, is employed for this purpose. Hence, the use of copper in agriculture is crucial. Another advanced and innovative application of copper is in chemical processes within the petrochemical industry as a catalyst. Copper catalysts exhibit a more favorable hydrogenation activity compared to nickel catalysts. The copper catalyst is designed in three forms: extruded and tablet forms for fixed-bed reactions and powder for liquid-bed reactions.

Keywords: copper catalyst, copper metal, copper alloys, copper applications, agricultural industry, electrical industries, pharmaceutical industries

1. Introduction

Copper metal is a natural element found in soil, water, and rocks. It is regarded as one of the functional metals that have significantly impacted the quality of human life. Historically, copper has been considered among the most essential materials in various industries [1]. Its extensive use in daily life, electronic industries, and industrial machinery underscores the paramount importance of this metal. Furthermore, copper stands out as one of the most widely used heavy nonferrous metals [2, 3]. In terms of appearance, copper compound minerals exhibit vibrant colors, especially red, green, and blue, as depicted in **Figure 1** [4].

Copper is a metal as old as human history. Copper was among the first metals used for coinage, a practice that started around 8000 BC. The coin depicted at the bottom is a Roman Follis featuring Constantius I (**Figure 2**) [5].

1.1 Characteristics and applications of copper

Why is copper a popular metal? Due to its exceptional and noteworthy features, copper is used in various industries such as machinery, agriculture, electronics,

Figure 1.
Copper ore.

Figure 2.
Roman Follis.

etc. [6]. Among the most essential features of copper metal, the following can be mentioned: conduction of electricity and heat [7], being a hammerhead, corrosion resistance [8], machining ability, tensile strength [9], alloy property [10], very high malleability, the possibility of recycling copper [11], and an esthetically pleasing color suitable for artistic applications.

The combination of these qualities has rendered copper a sought-after metal for various purposes. Furthermore, Copper, as a very versatile metal, has excellent electrical and thermal conductivity, malleability, corrosion resistance, and other desirable properties as a final product in several sectors such as (1) Electronics industries: production, transmission, and distribution of electricity. (2) Construction and architecture: roof and plumbing. (3) Transportation: car wires, connections, and various electrical parts for the construction of ships, planes, and trains. (4) Industrial machinery

Figure 3.
Utilization of copper metal.

and equipment: engines, generators, pumps, compressors, and heat exchangers. (5) Renewable energy: It is an essential component in solar panels, wind turbines, and energy storage systems. (6) Plumbing and water systems: plumbing and water supply systems. (7) Coins and currency: production of coins, especially with lower denominations. (8) Medical applications: Medical devices and equipment, including surgical instruments, imaging systems, touch surfaces, and hospital equipment are suitable to help reduce the spread of infection. (9) Consumer goods: cooking utensils, household appliances, jewelry, decorative items, and musical instruments are used. In other words, the direct sale of copper sheets or various forms is a common practice (**Figure 3**).

1.2 Role of copper alloys in industry

Copper metal alloys are crucial, especially brass, bronze, copper, and nickel silver. Approximately 30% of the world's copper production is dedicated to alloying. Copper alloys are generally worked, with only 10% is produced by casting, approximately 78–80% copper and 20–22% tin mixed together to form bell metal; in smaller proportions, it creates bronze. Additionally, 70–90% copper metal is mixed with 10–30% zinc to form a brass alloy, also 88% copper and 12% zinc used to form Pinchbeck, and other alloys (**Figure 4**) [10, 12, 13].

1.3 End uses of copper

End uses of copper metal and the use of copper products in various industries are called "end uses." In order of importance, these industries are electrical and electronic, construction, consumer goods, industrial machinery manufacturing, transportation, and military. Copper is one of the four metals whose natural color is not gray or silver. The most production and extraction of copper belongs to Chile. The sale of copper in Chile, with an exhibition of 7.5 million tons, has taken the largest share of the copper market. Copper ingot producers, foundries, wire-making units, brass products production units, and other sectors use refined copper. The transformation of copper into products such as uncoated wire, cable, coated wire, tape and belt, pipe, rod, and casting products reveals its final applications in our lives (**Figure 5**) [14].

Figure 4.
Copper alloys.

Figure 5.
Use in electrical industries.

2. Copper's versatility in daily life and industry

In the previous sections, we learned about the characteristics of copper metal. These features have increased the use of copper in different conditions. For this reason, today, we see the benefit of copper in various industries and daily life.

2.1 Application of copper in electronics and electrical industries

One of the best and most cost-effective conductive metals is copper, known for its efficient electrical conductivity. Additionally, owing to its properties, copper has become one of the safest metals for wires and power transmission tools. The majority

Figure 6.
The use of copper in the electronics industry.

of copper is utilized in the electrical and electronic industries, accounting for approximately 46% (**Figure 6**) [15].

One of the high-tech applications of copper in electrical applications is the use of copper nanoparticles for conductive inks. The potential of low price, low melting temperature, and high conductivity of copper nanoparticles has enabled them to prepare themselves to enter new flexible electronic technologies, such as foldable and wearable electronic devices (**Figure 7**) [16].

The new technologies, such as solar panels, wind turbines, and energy storage systems for renewable energy production, are experiencing growth due to the escalating consumption of fossil fuels and the resultant increase in pollution. These technologies, including solar, tidal, and wind, necessitate immediate storage solutions. Another high-tech electronic application is supercapacitors, which function as devices capable of rapidly storing and generating electrical energy. CuS/C@PANI nanocomposites are employed as advanced electrode materials for supercapacitor applications. These superconductors are composed prepared of copper nitrate trihydrate ($Cu(NO_3)_2{\cdot}3H_2O$, 99.99%), polyvinyl pyrrolidone (PVP, 99%), and 1,3,5-benzenetricarboxylic acid (H3BTC, 95%) (**Figure 8**) [17].

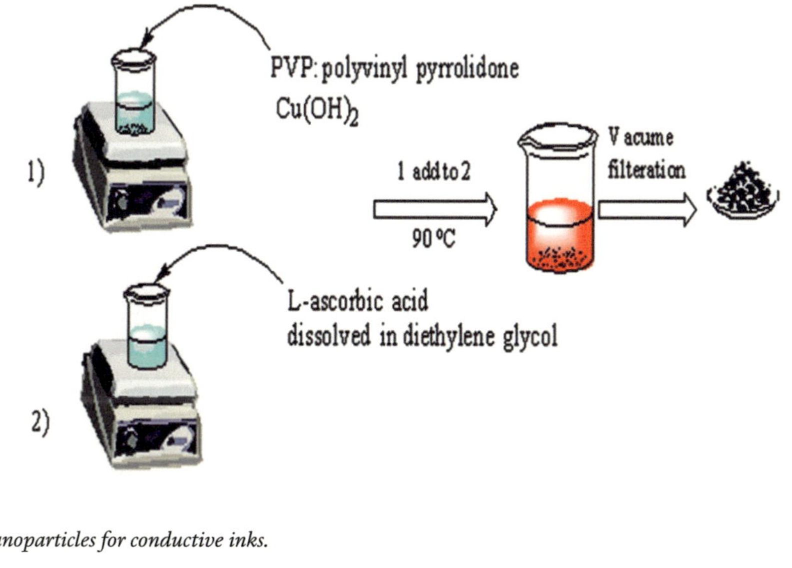

Figure 7.
Copper nanoparticles for conductive inks.

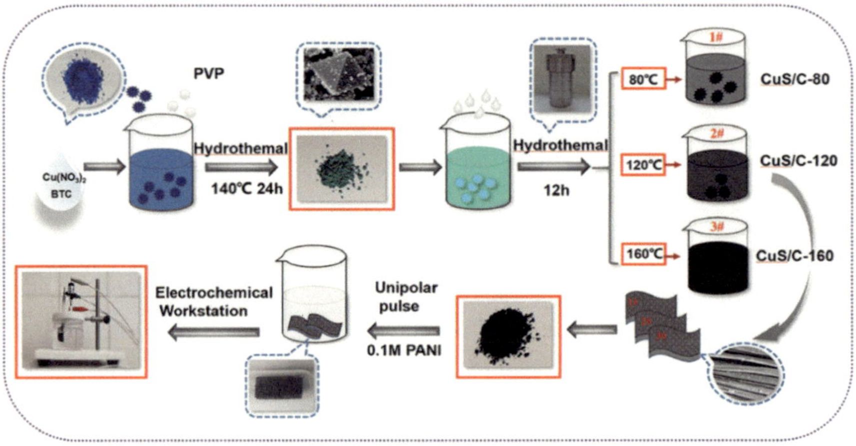

Figure 8.
Synthesis CuS/C@PANI nanocomposites as advanced electrode materials for supercapacitor applications [17].

2.2 Construction application

Copper metal finds application not only in the façade of buildings but also in all wiring, certain pipes, various door handles, and even in all the electrical appliances inside a house. Interestingly, almost 197 kg of copper is used in each building [18, 19]. One copper compound with unique applications is copper nanoparticles featuring different structures tailored for specific uses. These nanoparticles, produced by biological methods, including copper compounds and plant extracts, can have basic applications in producing paints, ceramics, batteries, fuel cells, etc. Bimetallic nanoparticles are compounds that can have these applications. Chitosan is a biological compound, biodegradable, biocompatible, and nontoxic polymer with antifungal properties, and a special chemical structure has a special application in the production of metal nanostructures, especially copper; due to the OH and NH2 groups, it can create a strong metal bond and create a complex and a strong polymer structure for special applications. It is employed as a stabilizing agent due to its ability to form chemical bonds with metals in the nanoparticle production process (**Figure 9**) [20].

2.3 Transportation industries application

Using copper to construct ships, railways, airplanes, and cars is also widespread in transportation industries. Copper alloys are standard materials in shipbuilding from bolts and rivets to a ship's propellers and pipes. This metal makes many train parts in the railway industry, including engines, brakes, and control parts. Airplanes require ventilation, hydraulic, navigation, and electrical systems, and copper is essential in the automotive industry for brakes, bearings, engines, radiators, and wiring. A typical vehicle alone can contain 23 kg of copper [21–23]. Copper oxide brake nanofluid (CBN) is a compound under consideration for car braking, evaluated using the arc submerged nanoparticle synthesis system (ASNSS). Brake fluids containing copper nanoparticles are produced by melting the bulk metal immersed in the dielectric liquid as an electrode. Copper vaporizes in DOT3 brake fluid, rapidly quenches, and results in the formation of nanocrystalline copper powders at the core. The produced CBN, aimed at reducing the occurrence of vapor lock, exhibits a higher boiling point,

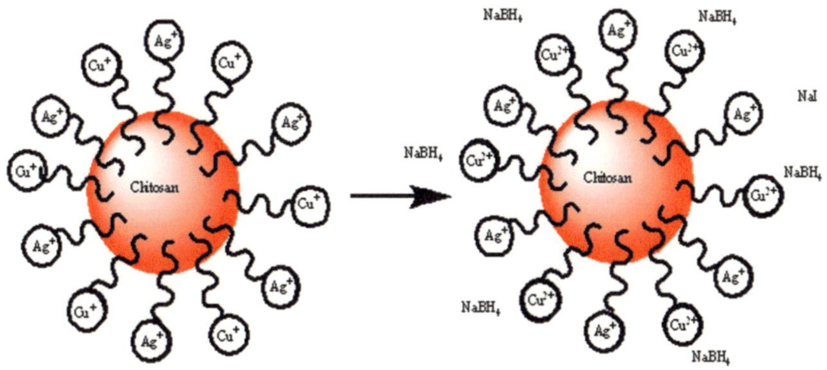

Figure 9.
Synthesis of bimetallic nanoparticles.

increased conductivity, and greater viscosity, ensuring the optimal performance of CBN. Considering that the performance of brake oil depends on the three characteristics of higher boiling point, higher viscosity, and stronger conductivity, one of the great advantages of CBN produced can be the increase of 8°C in the boiling point of CBN, which has reached 278°C. Also, an increase of 5 units of viscosity and reaching the viscosity at 35 (mm^2/s, cSt) to 37 (mm^2/s, cSt) and an increase of 0.05 W/m°C is 0.03 units of conductivity and reaching a conductivity of 0.05 W/m°C [24].

2.4 Pharmaceutical industries application

While copper's utilization in the pharmaceutical industry may not be as extensive as in other sectors, its significance persists owing to the unique properties of copper metal, notably its antibacterial characteristics. Studies have shown that bacteria, viruses, and yeast find it challenging to survive on a copper surface due to the metal's interference with the electrical charge on the microbial cell membrane. Among the bacteria and microorganisms in question are E. coli, Legionella pneumophila, cobacillus, Staphylococcus aureus, Salmonella, poliovirus cited. According to the US Environmental Protection Agency (U.S. EPA), a copper surface can eliminate 99.9% of bacteria that land on it within two hours [25–27]. The utilization of a composite film, demonstrating outstanding bacterial-killing efficiency, as a wound dressing material is one of the applications of copper metal as an antibacterial compound. In this process, copper nanoparticles with oxidative stability are produced using a modified dialdehyde starch (DAS) as an environmentally friendly reducing agent. Additionally, a biocompatible polymer, polyethyleneimine (PEI), serves as a stabilizer in the production of these copper nanoparticles. The potential for various clinical applications has been investigated using the antibacterial properties of agar film embedded with copper nanoparticles against E. coli bacteria, which has shown excellent bacterial killing efficiency and can be used as a wound dressing material (**Figure 10**) [28].

Healthy living conditions and international concerns have given rise to new challenges in developing cost-effective antimicrobial materials, focusing on preventing the survival and proliferation of microorganisms, particularly after the attention garnered by the coronavirus pandemic. An emerging field involves the utilization of metal cations in conjunction with azo-Schiff base ligands as highly active antibacterial compounds (**Figure 11**). Recent studies emphasize the noteworthy antimicrobial and antibacterial properties of copper metal ions on Staphylococcus aureus and E. coli bacteria [29].

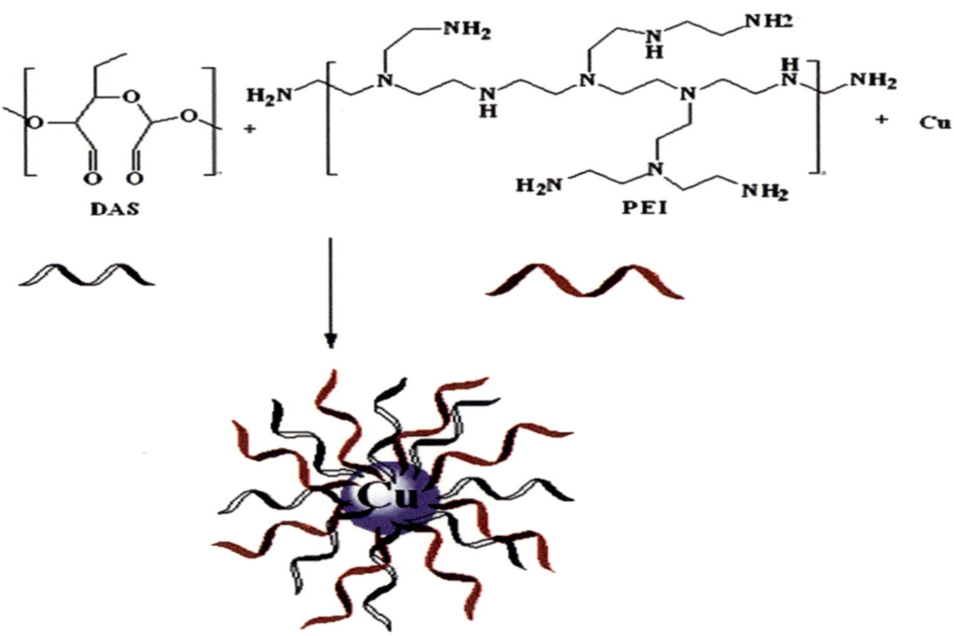

Figure 10.
Copper nanoparticles antibacterial compound.

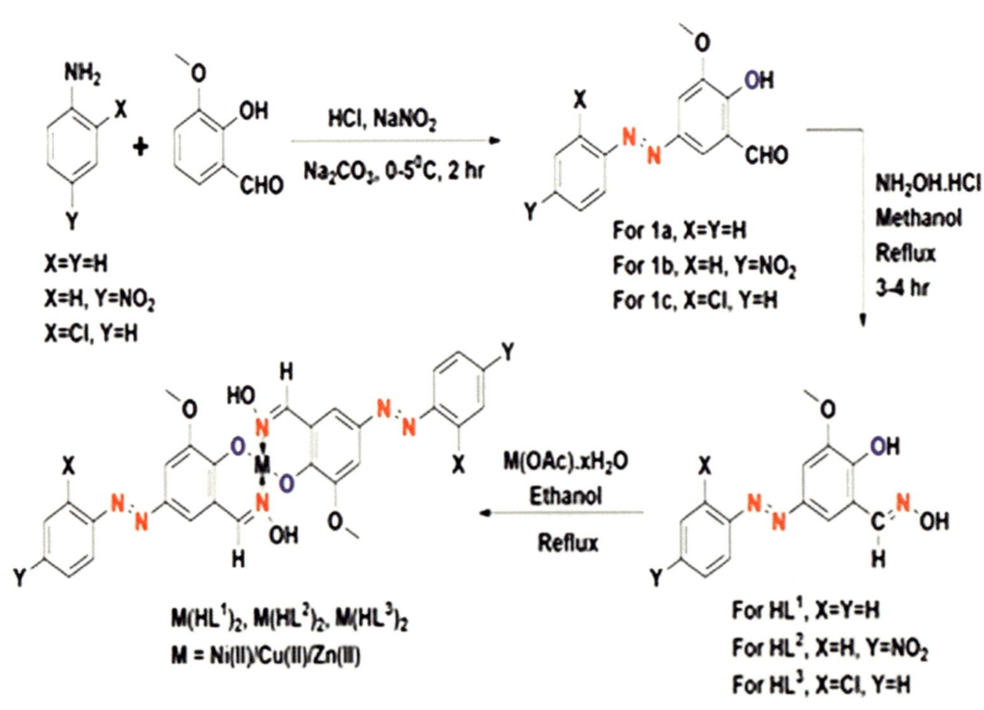

Figure 11.
Schematic representation for the synthesis route of azo-Schiff base ligands and their metal (II) complexes. Reprinted with permission from [29]. License number: 5677770865497.

2.5 The use of copper in industrial equipment and machinery application

The application of copper in industrial equipment and machinery is widespread across various sectors, including the petrochemical industry. These machines and equipment comprise copper pipes, electric motors, evaporators, condensers, heat exchangers, valves, and containers designed for use in corrosive environments [30, 31].

Investigating the chemical nature of copper oxide (Cu_2O) and copper surfaces in the presence of H_2O and CO_2 at room temperature using X-ray photoelectron spectroscopy shows that today the use of copper in rotor and brush electrodes is considered in electric motors. Investigating the surface chemistry of copper surfaces used in electric motors on Cu_2O, Cu, and Zn/Cu characters at room temperature, in the presence of CO_2 gas, using X-ray photoelectron spectroscopy under constant temperature and pressure environmental conditions shows that clean copper can activate CO_2 to $CO_2^{-ä}$ species with a negative charge. At the same time, Cu_2O is inactive for CO_2 absorption at room temperature CO_2 can be converted to carbonate more in the presence of Cu. This action increases the electric current in electric motors (**Figure 12**), the XPS spectrum of Cu_2O in the presence of 0.1 torr CO_2. It shows at room temperature (**Figure 13**) [32].

2.6 Copper metal in the agricultural application

The application of copper metal in the agricultural industry is crucial as it serves as a primary nutrient necessary for the optimal growth of living tissues in plants and other organisms. Additionally, copper is utilized to control fungal diseases and act as a stimulant for the development of chickens and other poultry. Copper sulfate, derived from copper metal, is one of the extensively used compounds in this context. Hence, the use of copper in agriculture is deemed essential [33–37].

The goal of safeguarding more sustainable agricultural products has become a priority for farmers. In 2020, the Food and Agriculture Organization of the United Nations (FAO)

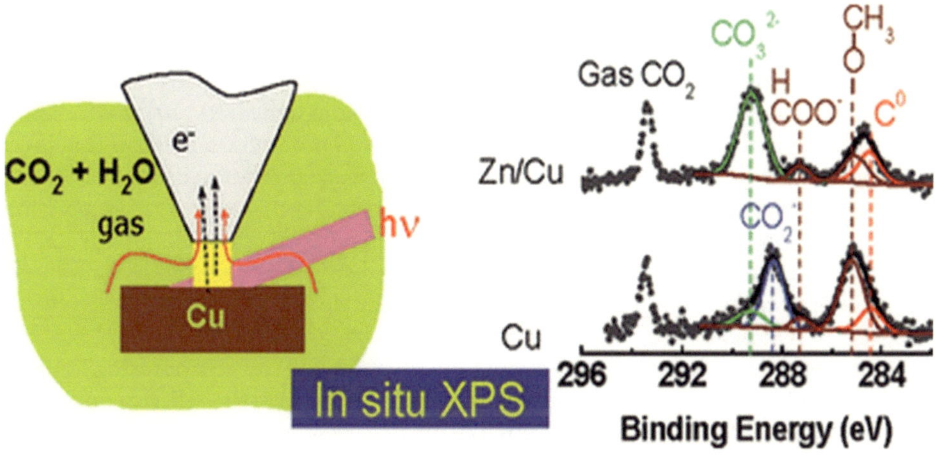

Figure 12.
Surface chemistry of Cu in the presence of H_2O and CO_2 [32]. Copyright © 2008, American Chemical Society.

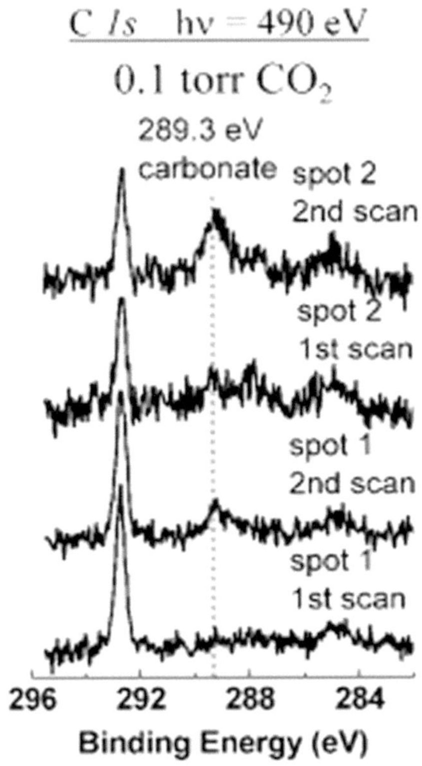

Figure 13.
C1s XPS spectra of Cu_2O in the presence of 0.1 torr CO_2 at room temperature, show that the X-rays and the secondary photoelectrons induce carbonate formation on Cu_2O [32]. Copyright © 2008, American Chemical Society.

designated it as an international year focused on protecting plants, promoting economic growth, and enhancing environmental sustainability. The FAO highlighted that over 40% of crops are lost annually due to pests. One historical copper-containing pesticide utilized for this purpose since 1885 is the Bordeaux mixture [38, 39]. It has been employed concurrently with pesticides, and subsequently, numerous pesticides incorporate copper compounds. The antibacterial activity of copper is grounded in two primary mechanisms. In the first mechanism, Cu^{2+} is generated and delivered [40]. In the second method, with the presence of copper nanoparticles, ROS production method that is reactive oxygen species proceeds [41]. Due to the decreased accumulation of metal in the soil, reduced harm to microbiota, and lower toxicity to plants, food, and underground water, the amount of copper in European pesticides has been reduced from 6 kg per hectare to 4 kg per hectare today [42]. Therefore, it is crucial to develop new pesticide compounds with low copper content and the most significant reduction of antimicrobial activity. Recent research indicates that lignin, an abundant, biodegradable, and natural biopolymer, can be used in conjunction with copper nanoparticles [43]. Incorporating lignin as polyphenols, which are antimicrobial compounds, can diminish the reliance on copper in the structure, while still imparting valuable properties, various phenolic monomer fragments are found in lignin, making it a valuable reservoir of natural antibacterial agents. The strongest inhibitory effect is generally observed in phenolic moieties with a double bond at the α and β positions of the side chain and a methyl group at the γ position (**Figure 14**) [45]. Antimicrobial properties have been successfully demonstrated in

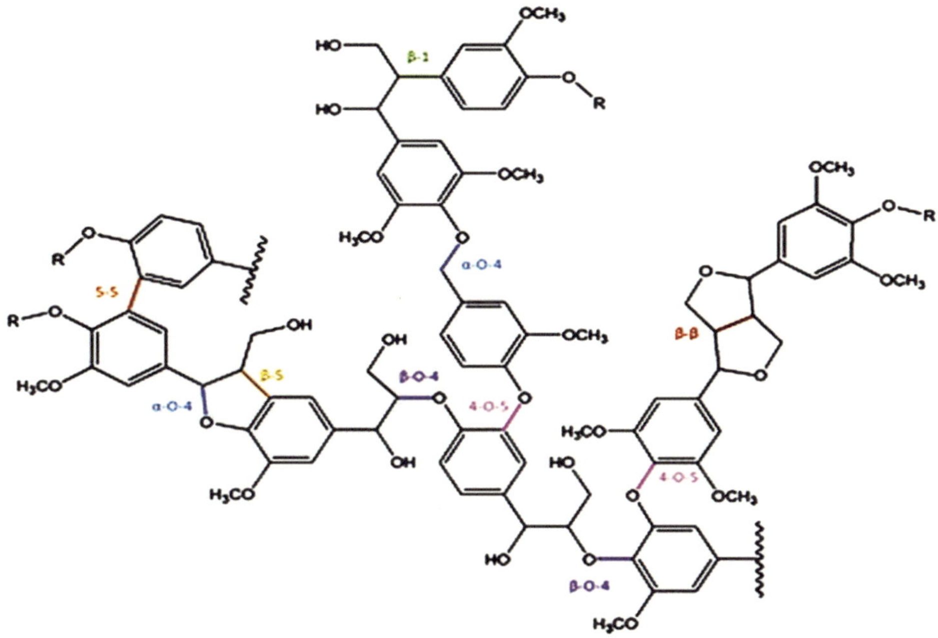

Figure 14.
Structure lignin as polyphenols [44] Copyright ©2022, American Chemical Society.

lignin derived from different sources, using different extraction methods, and even in a wide range of lignin model compounds against different microorganisms [46–48]. The antimicrobial effect of polyphenolic compounds of lignin is attributed to the damage and lysis of the bacterial cell membrane, which leads to the subsequent release of the cell contents [45, 49]. Nanoparticles show more useful polyphenolic side chains on their surface due to their wide surface area. This extended contact area has the potential to increase the antimicrobial effect [44]. An innovative organic-inorganic combination of lignin and brocanthite $Cu_4(OH)_6SO_4$ has garnered significant attention (**Figure 15**). In conclusion, the antimicrobial activity of the lignin@Cu compound and the correlation between the dimensions/shape of the brochantite crystals were investigated on the tomato plant through *in vivo* tests, yielding valuable results. The *in vivo* test revealed that the copper content in commercial pesticides is 20 times higher than that in the lignin@ Cu composition. Notably, the organic-inorganic compound with wood-shaped crystals measuring about 10–30 nm in length demonstrated significantly higher effectiveness compared to these conventional developments [50].

The use of metal ions in CuO, CuS, and $Cu(OH)_2$ compounds for antifungal properties is common in commercial applications. However, recent restrictions have emerged due to concerns about their accumulation in farm soil and water resource contamination, limiting their use. It is crucial to employ nanomaterials for controlling the release of copper metal [51]. A valuable method involves using phosphate compounds to regulate the release of metal ions and reduce fungal and bacterial infections. The compound $Cu_3(PO_4)_2{\cdot}3H_2O$, synthesized through the polyol method and presented by White and colleagues, has yielded valuable results in watermelon field applications to enhance resistance to fungi. A concentration of 10 mg L^{-1} nanosheet $Cu_3(PO_4)_2{\cdot}3H_2O$ exhibited a 58% reduction in the development of fungal diseases (**Figure 16**) [52].

Figure 15.
Two synthetic pathways followed to obtain the hybrid material lignin@Cu: (a) wet procedure, and (b) mechanochemistry [50]. Copyright ©2020, American Chemical Society.

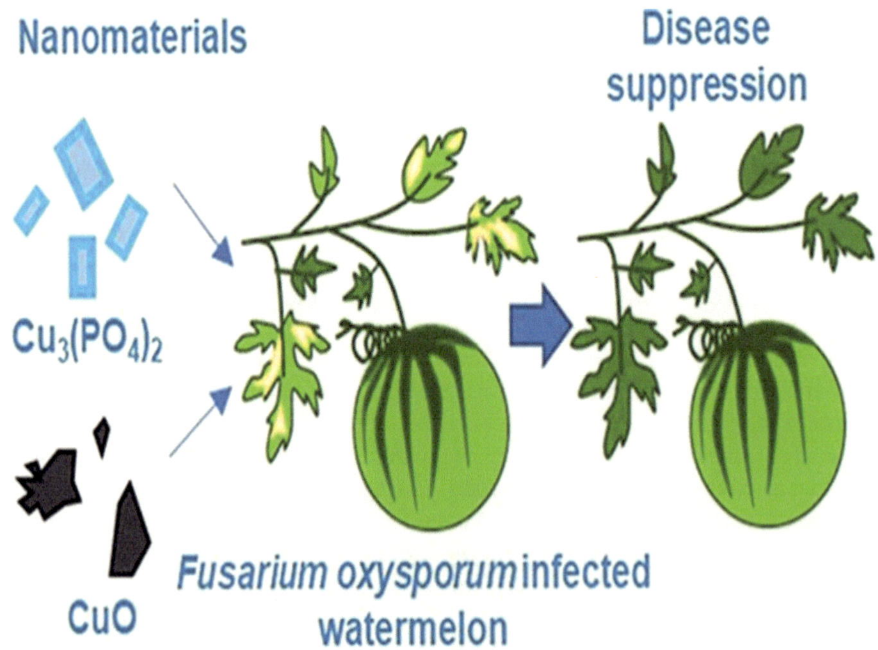

Figure 16.
Use of copper metal ions in watermelon field [52]. Copyright ©2020, American Chemical Society.

2.7 Copper catalyst applications

Due to their high potential applications, certain metals, including copper, are promising alternatives to expensive metals as catalysts in chemical processes. The natural abundance and low manufacturing cost of copper nanoparticles have positioned it as a valuable compound in synthesis [53]. Practical methods for synthesizing copper nanoparticles include sonochemical, hydrothermal, sol-gel, electrochemical, and thermal evaporation techniques. However, these methods often involve the use of regenerating chemical compounds, such as $NaBH_4$, or may require other substances, such as ethylene glycol and polyvinyl pyrrolidone, which can pose environmental risks [54]. Currently, the adoption of green synthesis methods for the production of metal nanoparticles is gaining prominence. Biological green synthesis, a well-established approach, utilizes natural biological resources, microorganisms, and plants. The stability of metal nanoparticles synthesized through biological methods is attributed to the presence of compounds such as flavonoids, alkaloids, and polyphenols. Compounds from the Rutaceae family, such as Muraya Koenigii (Curry leaf) or Daun Kari in Malay, are particularly prevalent in the biological synthesis of metal nanoparticles. Among the advantages of biological methods for the synthesis of copper nanoparticles, they offer mild reaction conditions, environmentally friendly, and cost-effective processes. These methods ensure biocompatibility and enable diverse applications in medicine while also providing control over the size and shape of the nanoparticles. The use of extracts or natural microorganisms reduces the environmental impact and energy consumption, making it a sustainable and efficient approach [55, 56]. Mustaffa Shamsuddin presented a valuable method using the aqueous extract of Murayya koenigii leaves as a reducing agent for producing CuO nanoparticles. He has used these nanoparticles as a catalyst in reducing 4-nitrophenol to 4-aminophenol (**Figure 17**) [57].

The development of green, efficient, solvent-free, and safe catalysts using one-pot methods for synthesizing heterocyclic compounds is a current focus. One of the green synthesis methods for heterocyclic compounds involves employing multicomponent reaction (MCR) [58]. This method holds significance in modern research for synthesizing heterocycles. It involves a technique where three or four starting materials combine to form a complex molecule through a one-pot reaction [59]. There are numerous methods for preparing heteroatoms containing (N and S) groups, such as pyrazole and thiadiazole [60]. However, the use and development of new catalytic methods with the design of nanocatalysts and features such as selectivity, biocompatibility, and reusability can be used as green industrial methods in preparing organic compounds [61]. One of the green synthesis methods for organic compounds involves the use of copper metal nanocatalysts [62]. Sharaf Zeebaree and his colleagues have explored the utilization of Trifolium resupinate leaf extract as a reducing agent for synthesizing copper nanoparticles, serving as catalysts in C−C and C-heteroatom bond formations. They have successfully synthesized a valuable heterocyclic compound known as 1,3,4-thiadiazole using the MCR method and the developed nanocatalyst. The synthesis method is illustrated in **Figure 18** [63].

A long-standing goal in polymer chemistry is the design of new macromolecules and achieving high molecular weight control using radical polymerization methods.

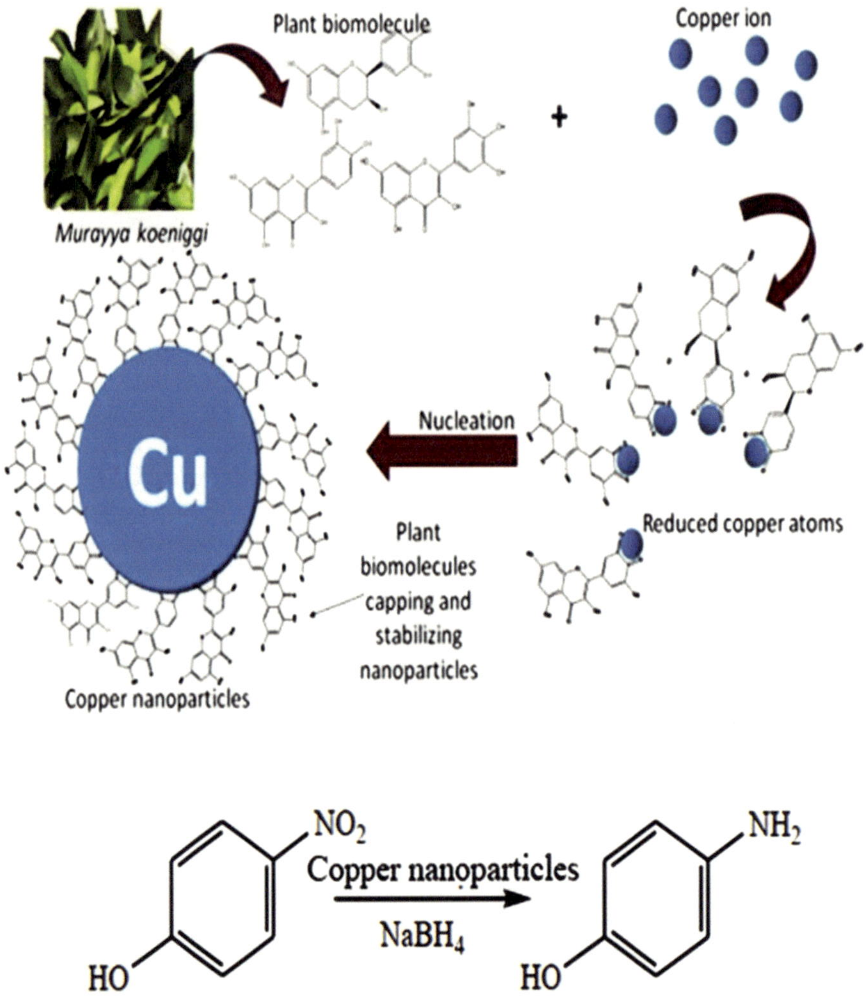

Figure 17.
Synthesis of copper nanoparticles as a catalyst for the reduction of 4-nitrophenol to 4-aminophenol [57].

In 1990, reversible deactivation radical polymerization (RDRP) techniques were first realized, as depicted in **Figure 19**. This figure illustrates the mechanism of atom transfer radical polymerization (ATRP), a type of RDRP reaction, with potential side reactions and equilibria in aqueous media. Generally, this method operates through a dynamic equilibrium between a dormant species and an actively propagating species. The conversion between inactive and active states occurs rapidly, with only a small fraction of chains being active and growing at any given moment. Consequently, the average molecular weight is influenced by the ratio between the initiator and monomer, ensuring that each polymer chain grows with the same molecular weight distribution [65].

The use of water as a reaction medium has transformed this process into a green one. The control and stability of vinyl monomer polymerization are other valuable features of the processes. New protocols for RDRP processes were introduced by Haddleton et al. in 2013, in which Cu(0) is produced in aqueous media by an *in situ*

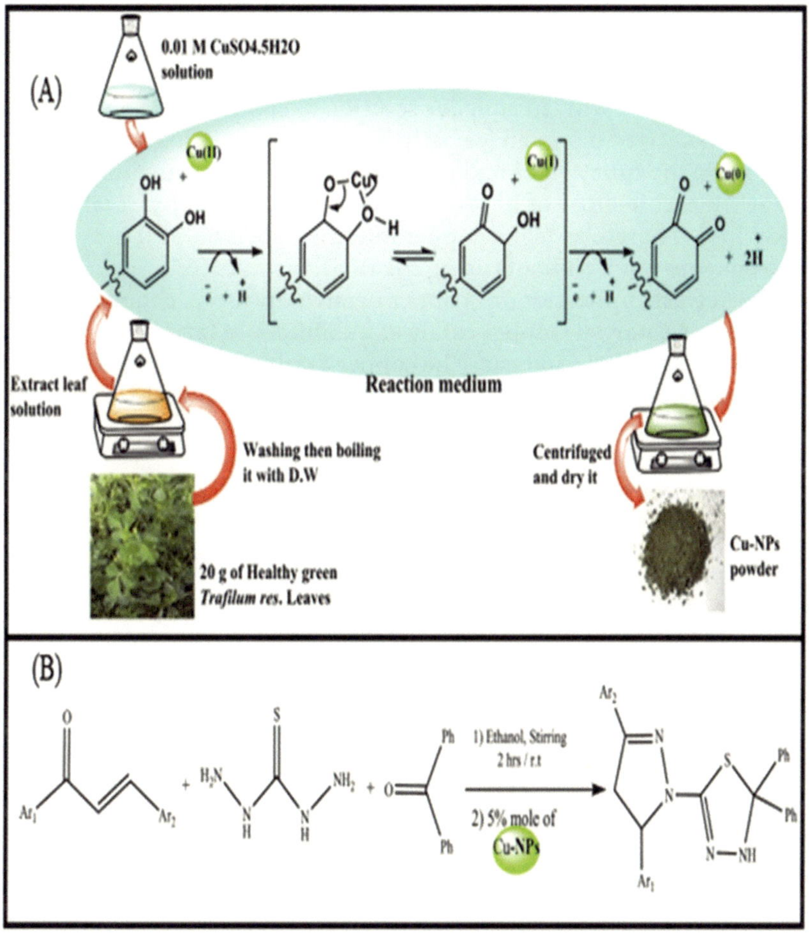

Figure 18.
(A) Synthesize copper nanoparticles and (B) synthesis of 1,3,4-thiadiazole [63]. License number: 5677750744431.

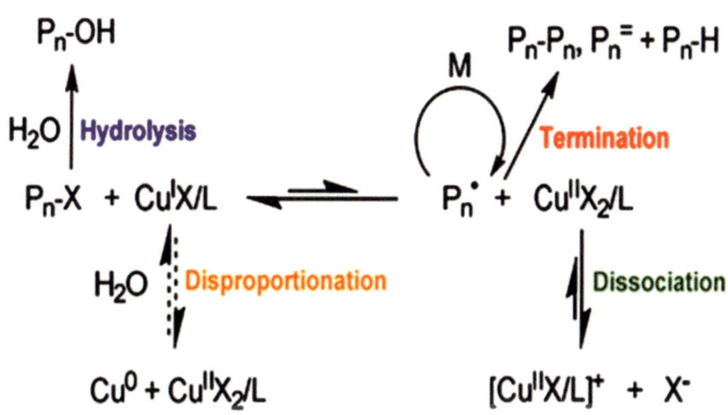

Figure 19.
Mechanism of ATRP, one of the RDRP reactions, with potential side reactions and equilibria in aqueous media: blue represents the hydrolysis of alkyl halide chain end, red depicts radical-radical termination reactions, orange illustrates the disproportionation of Cu(I), and green signifies the dissociation of halogens. Reprinted with permission from reference [64]. Copyright 2015 Royal Society of Chemistry.

production method. Cu(0) is formed through the rapid disproportionation of Cu(I) before the addition of monomers and initiators [66]. Then, using nitrogen injection, deoxygenation is created in the solution. Monomers and alkyl halides are injected into the catalyst mixture with an aqueous solution to complete the polymerization. Polymerization is carried out in an ice bath due to a decrease in the rate of hydrolysis and an increase in the function of end groups (**Figure 20**). Alsubaie and his colleagues made the architecture of high-grade block copolymers by decablock copolymers quickly [64]. Therefore, using copper intermediate is helpful in synthesizing polymers in aqueous environments using the RDRP process [67].

One of the newest applications of copper involves its use in chemical processes in petrochemicals as a catalyst. Copper catalysts exhibit more favorable hydrogenation activity compared to nickel catalysts. The copper catalyst is designed in three forms: extruded and tablet for fixed-bed reactions and powder for liquid-bed reactions.

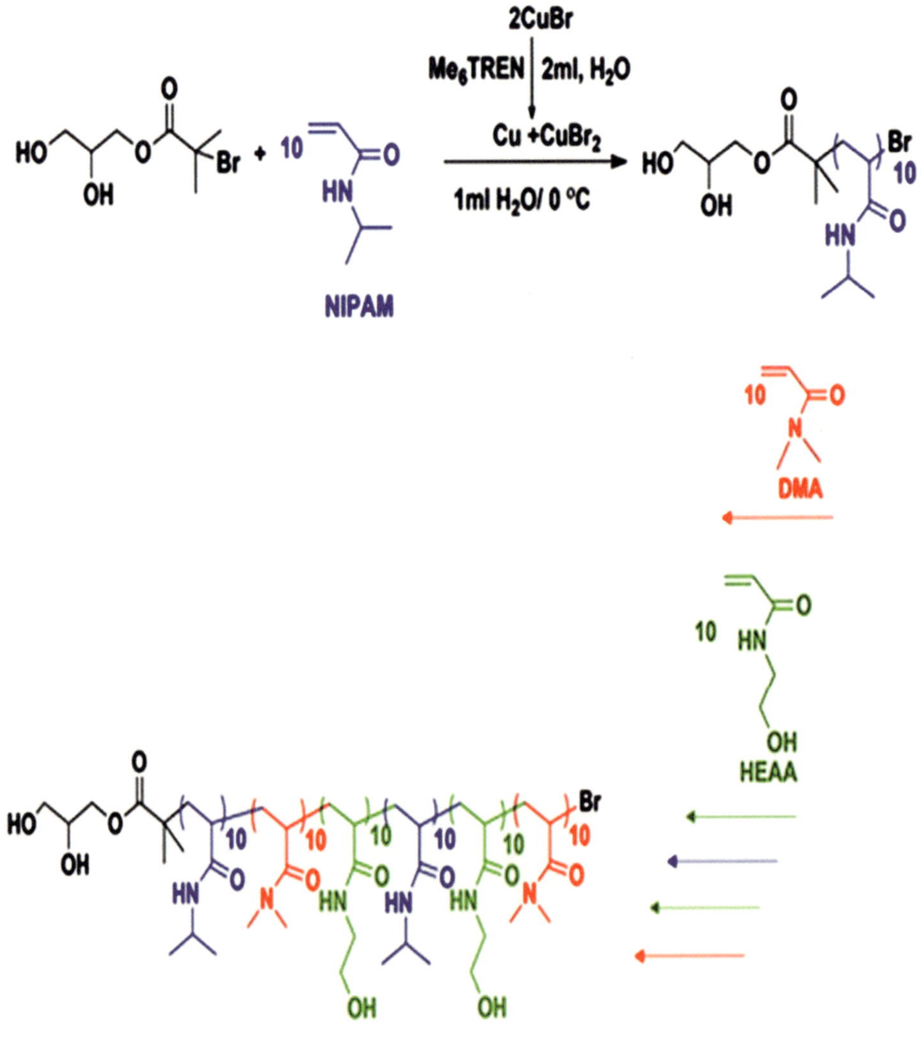

Figure 20.
Synthesis of multi-block copolymers composed of NiPAm, DMA, and HEAm by iterative Cu(0)-RDRP in H_2O, permission from reference [64]. Copyright 2015 Royal Society of Chemistry.

A crucial chemical transformation in the industry, representing one of the applications of gas-phase purification in the flow of olefins for polymerization, is the semi-hydrogenation of alkyne to alkene [68, 69]. The synthesis of metal nanoparticles with high-added value is widely utilized in the industry. In 2022, Mohammadi Dehcheshmeh et al. conducted a green process to synthesize nanoparticles of palladium and copper. It has had an exciting performance in the hydrogenation process. PdCu nanoparticles are synthesized through three steps, including I- synthesis of polylactide as a biodegradable and widely used polymer II- binding of polylactide to metal and formation of metal stereocomplexes based on polylactide III- reduction of metal using hydrogen (**Figure 21**) [70].

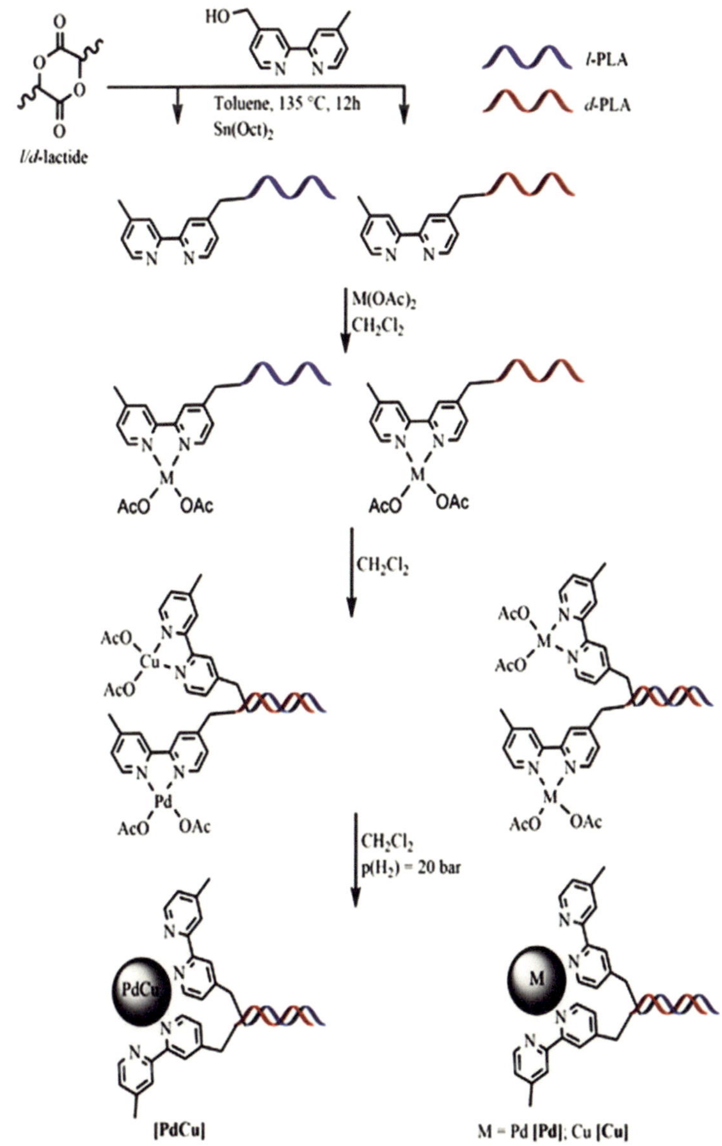

Figure 21.
Syntheses of the nanoparticles catalysts PdCu, Pd, and Cu [70]. License number: 5680370589458.

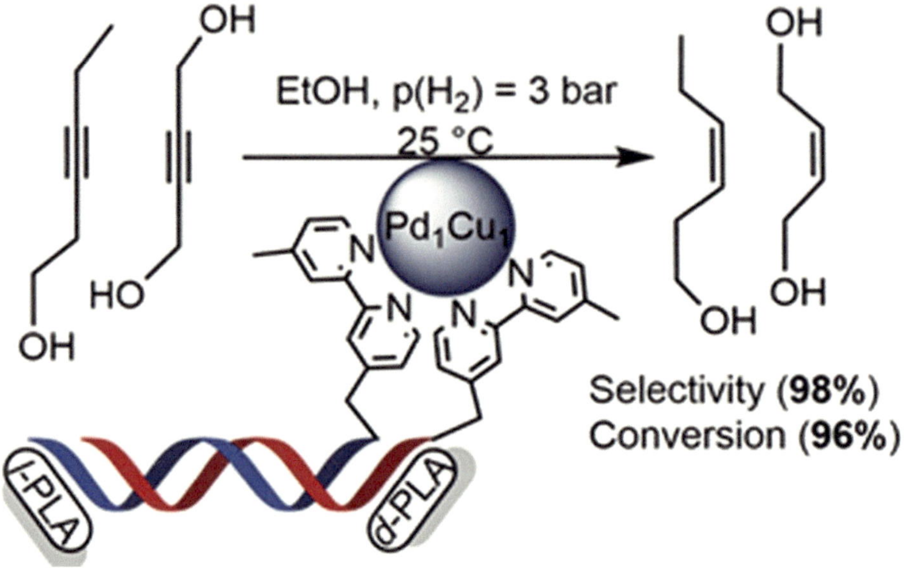

Figure 22.
Hydrogenation by Cu nanoparticles catalyst [70]. License number: 5680370589458.

One of the applications of these nanoparticles is the partial hydrogenation of 2-butene-1,4-diol and 3-hexyn-1-ol as an industrially important alkynol, which led to cis-alkenol with high selectivity (98%) (**Figure 22**).

3. Conclusion

This chapter emphasizes the significance of utilizing copper as a versatile metal applicable in pharmaceutical, electronic, chemical, industrial, and agricultural domains. The use of nanoparticles represents a scientific revolution in the twenty-first century, with copper nanoparticles playing a pivotal role in various industries. These nanoparticles serve as fundamental elements in fields such as agriculture, pharmaceuticals, electronics, and industry, functioning as catalysts. Consequently, the green synthesis methods outlined in this chapter hold particular importance. The green synthesis of copper nanoparticles has garnered attention from scientists and researchers as an advanced technology.

DOI: http://dx.doi.org/10.5772/intechopen.1004068

Author details

Iman Mohammadi Dehcheshmeh[1*], Ahmad Poursattar Marjani[1], Fatemeh Sadegh[2] and Mohammad Ebrahim Soltani[3]

1 Faculty of Chemistry, Department of Organic Chemistry, Urmia University, Urmia, Iran

2 Faculty of Sciences, Department of Chemistry, University of Sistan and Baluchestan, Zahedan, Iran

3 Chemical Engineering Department, Isfahan University of Technology, Isfahan, Iran

*Address all correspondence to: imanmohammadi88@gmail.com

References

[1] Baker DE, Amacher MC. Nickel, copper, zinc, and cadmium. In: Methods of Soil Analysis: Part 2 Chemical and Microbiological Properties. American Society of Agronomy, Inc., Soil Science Society of America, Inc.; 1983. pp. 323-336. DOI: 10.2134/agronmonogr9.2.2ed.c19

[2] Boulamanti A, Moya JA. Production costs of the non-ferrous metals in the EU and other countries: Copper and zinc. Resources Policy. 2016;**49**:112-118. DOI: 10.1016/j.resourpol.2016.04.011

[3] Darling A. Non-ferrous metals. In: An Encyclopaedia of the History of Technology. Google Book; 2002. pp. 47-145

[4] Behar O, Peña R, Kouro S, Kracht W, Fuentealba E, Moran L, et al. The use of solar energy in the copper mining processes: A comprehensive review. Cleaner Engineering and Technology. 2021;**4**:100259. DOI: 10.1016/j.clet.2021.100259

[5] Johanson P. Copper. The Rosen Publishing Group. Inc. Google Book; 2006

[6] Bhagat M, Anand R, Sharma P, Rajput P, Sharma N, Singh K. Multifunctional copper nanoparticles: Synthesis and applications. ECS Journal of Solid State Science and Technology. 2021;**10**:063011. DOI: 10.1149/2162-8777/ac07f8

[7] Liu H, Shi X, Xu F, Zhang L, Zhang W, Chen L, et al. Copper ion liquid-like thermoelectrics. Nature Materials. 2012;**11**:422-425. DOI: 10.1038/nmat3273

[8] Brusic V, Frisch M, Eldridge B, Novak F, Kaufman F, Rush B, et al. Copper corrosion with and without inhibitors. Journal of the Electrochemical Society. 1991;**138**:2253. DOI: 10.1149/1.2085957

[9] Merz M, Dahlgren S. Tensile strength and work hardening of ultrafine-grained high-purity copper. Journal of Applied Physics. 1975;**46**:3235-3237. DOI: 10.1063/1.322048

[10] Davis JR. Copper and Copper Alloys. ASM International. Google Book; 2001

[11] Samuelsson C, Björkman B. Copper recycling. In: Handbook of Recycling. New York: State-of-the-art for Practitioners, Analysts, and Scientists; 2014. pp. 85-94. DOI: 10.1016/B978-0-12-396459-5.00007-6

[12] Won C-S, Jung JP, Won K-S, Sharma A. Technological insights into the evolution of bronze bell metal casting on the Korean Peninsula. Metals. 2022;**12**:1776. DOI: 10.3390/met12111776

[13] Enguita O, Climent-Font A, Garcıa G, Montero I, Fedi M, Chiari M, et al. Characterization of metal threads using differential PIXE analysis. Nuclear Instruments and Methods in Physics Research B. 2002;**189**:328-333. DOI: 10.1016/S0168-583X(01)01081-3

[14] Reolon LW, Henning Laurindo CA, Torres RD, Amorim FL. WEDM performance and surface integrity of Inconel alloy IN718 with coated and uncoated wires. International Journal of Advanced Manufacturing Technology. 2019;**100**:1981-1991. DOI: 10.1007/s00170-018-2828-6

[15] Leisner P, Möller P, Fredenberg M, Belov I. Recent progress in pulse reversal plating of copper for electronics

applications. Transactions of the IMF. 2007;**85**:40-45. DOI: 10.1179/174591907X161973

[16] Cheng C, Li J, Shi T, Yu X, Fan J, Liao G, et al. A novel method of synthesizing antioxidative copper nanoparticles for high performance conductive ink. Journal of Materials Science: Materials in Electronics. 2017;**28**:13556-13564. DOI: 10.1007/s10854-017-7195-9

[17] Liu Q, Zhang S, Xu Y. Two-step synthesis of CuS/C@ PANI nanocomposite as advanced electrode materials for supercapacitor applications. Nanomaterials. 2020;**10**:1034. DOI: 10.3390/nano10061034

[18] Shah KW, Lu Y. Morphology, large scale synthesis and building applications of copper nanomaterials. Construction and Building Materials. 2018;**180**:544-578. DOI: 10.1016/j.conbuildmat.2018.05.159

[19] Madheswaran C, Ambily P, Dattatreya J, Rajamane N. Studies on use of copper slag as replacement material for river sand in building constructions. Journal of The Institution of Engineers (India): Series A. 2014;**95**:169-177. DOI: 10.1007/s40030-014-0084-9

[20] Din MI, Rehan R. Synthesis, characterization, and applications of copper nanoparticles. Analytical Letters. 2017;**50**:50-62. DOI: 10.1080/00032719.2016.1172081

[21] Lavanya C, Rao AS, Kumar ND. A review on utilization of copper slag in geotechnical applications. In: Indian Geotechnical Conference. Kochi. Paper No.H-212. December; 2011, India. pp. 445-448

[22] Crowson P. The copper industry 1945-1975. Resources Policy. 2007;**32**:1-18. DOI: 10.1016/j.resourpol.2007.03.004

[23] Lamblin G, Duhain A, Lenoble D. Copper-carbon nanotubes composite materials for lightning strike protections in airplanes. ECS Meeting Abstracts. 2021;**7**:1891-1891. DOI: 10.1149/MA2021-0271891mtgabs

[24] Kao M, Lo C, Tsung T, Wu Y, Jwo C, Lin H. Copper-oxide brake nanofluid manufactured using arc-submerged nanoparticle synthesis system. Journal of Alloys and Compounds. 2007;**434**:672-674. DOI: 10.1016/j.jallcom.2006.08.305

[25] Konieczny J, Rdzawski Z. Antibacterial properties of copper and its alloys. Archives of Materials Science and Engineering. 2012;**56**:53-60

[26] Edis Z, Haj Bloukh S, Ashames A, Ibrahim M. Copper-based nanoparticles, their chemistry and antibacterial properties: A review. Chemistry for a Clean and Healthy Planet. 2019;**1**:401-428. DOI: 10.1007/978-3-030-20283-5_24

[27] Zhang E, Li F, Wang H, Liu J, Wang C, Li M, et al. A new antibacterial titanium–copper sintered alloy: Preparation and antibacterial property. Materials Science and Engineering: C. 2013;**33**:4280-4287. DOI: 10.1016/j.msec.2013.06.016

[28] Tang L, Zhu L, Tang F, Yao C, Wang J, Li L. Mild synthesis of copper nanoparticles with enhanced oxidative stability and their application in antibacterial films. Langmuir. 2018;**34**:14570-14576. DOI: 10.1021/acs.langmuir.8b02470

[29] Kasare MS, Dhavan PP, Jadhav BL, Pawar SD. Synthesis of Azo Schiff Base ligands and their Ni (II), Cu (II) and Zn (II) metal complexes as highly-active

antibacterial agents. ChemistrySelect. 2019;**4**:10792-10797. DOI: 10.1002/slct.201901605

[30] MacMahon AD. Copper: A Materials Survey. US Department of the Interior. Bureau of Mines. Google Book; 1965

[31] Foote KG. Optimizing copper spheres for precision calibration of hydroacoustic equipment. The Journal of the Acoustical Society of America. 1982;**71**:742-747. DOI: 10.1121/1.387497

[32] Deng X, Verdaguer A, Herranz T, Weis C, Bluhm H, Salmeron M. Surface chemistry of Cu in the presence of CO_2 and H_2O. Langmuir. 2008;**24**:9474-9478. DOI: 10.1021/la8011052

[33] Borkow G, Gabbay J. Copper, an ancient remedy returning to fight microbial, fungal and viral infections. Current Chemical Biology. 2009;**3**:272-278. DOI: 10.2174/187231309789054887

[34] Kumar V, Pandita S, Sidhu GPS, Sharma A, Khanna K, Kaur P, et al. Copper bioavailability, uptake, toxicity and tolerance in plants: A comprehensive review. Chemosphere. 2021;**262**:127810. DOI: 10.1016/j.chemosphere.2020.127810

[35] Lamichhane JR, Osdaghi E, Behlau F, Köhl J, Jones JB, Aubertot J-N. Thirteen decades of antimicrobial copper compounds applied in agriculture. A review. Agronomy for Sustainable Development. 2018;**38**:1-18. DOI: 10.1007/s13593-018-0503-9

[36] Salam D, El-Fadel M. Mobility and availability of copper in agricultural soils irrigated from water treated with copper sulfate algaecide. Water, Air, and Soil Pollution. 2008;**195**:3-13. DOI: 10.1007/s11270-008-9722-z

[37] Richardson HW. Handbook of Copper Compounds and Applications. CRC Press. Google book; 1997

[38] FAO. The State of Food and Agriculture 2019. Moving Forward on Food Loss and Waste Reduction. Vol. 3. Rome: FAO; 2019. pp. 2-13

[39] Johnson GF. The early history of copper fungicides. Agricultural History. JSTOR. 1935;**9**(2):67-79

[40] Warnes S, Caves V, Keevil C. Mechanism of copper surface toxicity in Escherichia coli O157: H7 and salmonella involves immediate membrane depolarization followed by slower rate of DNA destruction which differs from that observed for gram-positive bacteria. Environmental Microbiology. 2012;**14**:1730-1743. DOI: 10.1111/j.1462-2920.2011.02677.x

[41] Slavin YN, Asnis J, Hńfeli UO, Bach H. Metal nanoparticles: Understanding the mechanisms behind antibacterial activity. Journal of Nanobiotechnology. 2017;**15**:1-20. DOI: 10.1186/s12951-017-0308-z

[42] Authority EFS, Arena M, Auteri D, Barmaz S, Bellisai G, Brancato A, et al. Peer review of the pesticide risk assessment of the active substance copper compounds copper (I), copper (II) variants namely copper hydroxide, copper oxychloride, tribasic copper sulfate, copper (I) oxide, Bordeaux mixture. EFSA Journal. 2018;**16**:e05152. DOI: 10.2903/j.efsa.2018.5152

[43] Hatakka A. Biodegradation of lignin. In: Hofrichter M, Steinbuchel A, editors. Biopolymers. Lignin, Humic Substances and Coal. Vol. 1. Weinheim: Wiley-VCH; 2005. pp. 129-180. DOI: 10.1002/3527600035.bpol1005

[44] Beisl S, Friedl A, Miltner A. Lignin from micro-to nanosize: applications. International Journal of Molecular Sciences. 2017;**18**:2367. DOI:10.3390/ijms18112367

[45] Cazacu G, Capraru M, Popa V. Advances in natural polymers. In: Thomas S, Visakh PM, Mathew AP, editors. Advanced Structured Materials. Berlin/Heidelberg, Germany: Springer; 2013

[46] Nada A, El-Diwany A, Elshafei A. Infrared and antimicrobial studies on different lignins. Acta Biotechnologica. 1989;**9**(3):295-298. DOI: 10.1002/abio.370090322

[47] Zemek J, Košíková B, Augustin J, Joniak D. Antibiotic properties of lignin components. Folia Microbiologica. 1979;**24**:483-486. DOI: 10.1007/BF02927180

[48] Sláviková E, Košíková B. Inhibitory effect of lignin by-products of pulping on yeast growth. Folia Microbiologica. 1994;**39**:241-243. DOI: 10.1007/BF02814656

[49] Telysheva G, Dizhbite T, Lebedeva G, Nikoleava V. Lignin products for decontamination of environment objects from pathogenic microorganisms and pollutants. In: Proceedings of the 7th ILI Forum, Barcelona, Spain. 2005

[50] Gazzurelli C, Migliori A, Mazzeo PP, Carcelli M, Pietarinen S, Leonardi G, et al. Making agriculture more sustainable: An environmentally friendly approach to the synthesis of lignin@ Cu pesticides. ACS Sustainable Chemistry & Engineering. 2020;**8**:14886-14895. DOI: 10.1021/acssuschemeng.0c04645

[51] Huang Y, Zhao L, Keller AA. Interactions, transformations, and bioavailability of nano-copper exposed to root exudates. Environmental Science & Technology. 2017;**51**:9774-9783. DOI: 10.1021/acs.est.7b02523

[52] Borgatta J, Ma C, Hudson-Smith N, Elmer W, Plaza Perez CD, De La Torre-Roche R, et al. Copper based nanomaterials suppress root fungal disease in watermelon (*Citrullus lanatus*): Role of particle morphology, composition and dissolution behavior. ACS Sustainable Chemistry & Engineering. 2018;**6**:14847-14856. DOI: 10.1021/acssuschemeng.8b03379

[53] Gawande MB, Goswami A, Felpin F-X, Asefa T, Huang X, Silva R, et al. Cu and Cu-based nanoparticles: Synthesis and applications in catalysis. Chemical Reviews. 2016;**11**:63722-63811. DOI: 10.1021/acs.chemrev.5b00482

[54] Tran TH, Nguyen VT. Copper oxide nanomaterials prepared by solution methods, some properties, and potential applications: A brief review. International Scholarly Research Notices. 2014;**2014**:56592. DOI: 10.1155/2014/856592

[55] Hasan M, Ullah I, Zulfiqar H, Naeem K, Iqbal A, Gul H, et al. Biological entities as chemical reactors for synthesis of nanomaterials: Progress, challenges and future perspective. Materials Today Chemistry. 2018;**8**:13-28. DOI: 10.1016/j.mtchem.2018.02.003

[56] Jain M, Gilhotra R, Singh RP, Mittal J. Curry leaf (*Murraya koenigii*): A spice with medicinal property. MOJ Biology and Medicine. 2017;**2**:236-256

[57] Nordin NR, Shamsuddin M. Biosynthesis of copper (II) oxide nanoparticles using Murayya koeniggi aqueous leaf extract and its catalytic activity in 4-nitrophenol reduction. Malaysian Journal of Fundamental and Applied Sciences. 2019;**15**:218-224. DOI: 10.11113/mjfas.v15n2.1390

[58] Bansal R, Soni PK, Ahirwar MK, Halve AK. One-pot multi-component synthesis of some pharmacologically significant 2, 4, 5-tri and 1, 2, 4,

5-Tetrasubstituted Imidazoles: A review. International Research Journal of Pure and Applied Chemistry. 2016;**11**:1-26. DOI: 10.9734/IRJPAC/2016/24493

[59] Brauch S, van Berkel SS, Westermann B. Higher-order multicomponent reactions: Beyond four reactants. Chemical Society Reviews. 2013;**42**:4948-4962. DOI: 10.1039/C3CS35505E

[60] Kumar KA, Jayaroopa P. Pyrazoles: Synthetic strategies and their pharmaceutical applications-an overview. International Journal of PharmTech Research. 2013;**5**:1473-1486

[61] Safaei-Ghomi J, Ghasemzadeh MA. An efficient multi-component synthesis of 14-aryl-14H-dibenzo [a, j] xanthene derivatives by AgI nanoparticles. Journal of Saudi Chemical Society. 2015;**19**:642-649. DOI: 10.1016/j.jscs.2012.05.007

[62] Inamdar SM, More VK, Mandal SK. CuO nano-particles supported on silica, a new catalyst for facile synthesis of benzimidazoles, benzothiazoles and benzoxazoles. Tetrahedron Letters. 2013;**54**:579-583. DOI: 10.1016/j.tetlet.2012.11.091

[63] Zeebaree SYS, Zeebaree AYS. Synthesis of copper nanoparticles as oxidising catalysts for multi-component reactions for synthesis of 1, 3, 4-thiadiazole derivatives at ambient temperature. Sustainable Chemistry and Pharmacy. 2019;**13**:100155. DOI: 10.1016/j.scp.2019.100155

[64] Alsubaie F, Anastasaki A, Wilson P, Haddleton DM. Sequence-controlled multi-block copolymerization of acrylamides via aqueous SET-LRP at 0 C. Polymer Chemistry. 2015;**6**:406-417. DOI: 10.1039/C4PY01066C

[65] Hailes HC. Reaction solvent selection: The potential of water as a solvent for organic transformations. Organic Process Research and Development. 2007;**11**:114-120. DOI: 10.1021/op060157x

[66] Zhang Q, Wilson P, Li Z, McHale R, Godfrey J, Anastasaki A, et al. Aqueous copper-mediated living polymerization: Exploiting rapid disproportionation of CuBr with Me6TREN. Journal of the American Chemical Society. 2013;**135**:7355-7363. DOI: 10.1021/ja4026402

[67] Jones GR, Anastasaki A, Whitfield R, Engelis N, Liarou E, Haddleton DM. Copper-mediated reversible deactivation radical polymerization in aqueous media. Angewandte Chemie, International Edition. 2018;**57**:10468-10482. DOI: 10.1002/anie.201802091

[68] Borodziński A, Bond GC. Selective hydrogenation of ethyne in ethene-rich streams on palladium catalysts. Part 1. Effect of changes to the catalyst during reaction. Catalysis Reviews. 2006;**48**:91-144. DOI: 10.1080/01614940500364909

[69] Eggersdorfer M, Laudert D, Létinois U, McClymont T, Medlock J, Netscher T, et al. Einhundert Jahre Vitamine–eine naturwissenschaftliche Erfolgsgeschichte. Angewandte Chemie. 2012;**124**:13134-13165. DOI: 10.1002/ange.201205886

[70] Oberhauser W, Frediani M, Mohammadi Dehcheshmeh I, Evangelisti C, Poggini L, Capozzoli L, et al. Selective alkyne semi-hydrogenation by PdCu nanoparticles immobilized on Stereocomplexed poly (lactic acid). ChemCatChem. 2022;**14**:e202101910. DOI: 10.1002/cctc.202101910

Section 5

Environmental Issues

Chapter 8

Advances in Techniques for Copper Analysis in Aqueous Systems

Ahmed Elkhatat

Abstract

Copper is an essential micronutrient but can be toxic at elevated levels. Monitoring copper in aqueous systems is critical for characterizing pollution sources and mitigating human health risks. This chapter comprehensively evaluates recent advances in analytical methods for detecting copper, including atomic spectrometry, molecular spectrophotometry, electrochemical sensors, voltammetry, and chromatography. Each technique's critical detection limits, selectivity, complexity, and advantages are outlined. Atomic absorption spectrometry, inductively coupled plasma-optical emission, and inductively coupled plasma-mass spectrometry provide the most sensitive copper quantification down to parts per trillion levels. Meanwhile, spectroscopic methods using novel reagents offer inexpensive and rapid copper screening. Electrochemical and optical sensors show promise for on-site and continuous monitoring. Chromatographic separation before detection improves selectivity in complex sample matrices. Critical evaluation of these complementary approaches can inform the selection of optimal copper quantification techniques for different environmental, industrial, and biological monitoring applications. Recent advances continue to expand the analytical toolkit for sensitive, selective, and cost-effective copper analysis across diverse aqueous systems.

Keywords: copper, water analysis, heavy metals, analytical methods, atomic spectrometry, sensors, voltammetry, chromatography

1. Introduction

The lack of clean and safe drinking water is a severe issue threatening human health and quality of life worldwide. The rapid increase in population and industrialization have led to a higher demand for drinkable water, resulting in increased wastewater contaminated with pollutants. Wastewater discharge significantly contributes to drinking water, groundwater, and marine environment pollution, especially with heavy metals. Although some heavy metals are necessary for biological systems, excessive exposure can harm ecosystems. For instance, exposure to heavy metals can result in bioaccumulation in fish's liver and muscle tissues, which can occur through different exposure pathways. These contaminants could severely affect marine life and human health if they enter the food chain. For instance, water pollution caused by heavy metals like lead, mercury, cadmium, chromium, copper, nickel, zinc, and arsenic poses a significant threat to our environment, leading to severe ecological and human health consequences. Heavy metals can enter water bodies through various pathways, including industrial wastewater discharges from mining, metal plating, electronics, and chemical manufacturing. Agricultural runoff, such as pesticides, fertilizers, and animal manure, can also contain heavy metals that may leach into water. Natural mineral deposits can erode and weather rocks and soil, releasing heavy metals. Corroded pipes and plumbing, particularly those made of lead and copper, can also leach into drinking water. Heavy metals released into the air from industries, incinerators, and so on can deposit via particulate matter into water. Leached heavy metals from landfills can enter surface and groundwater through rainfall percolating through the landfill. Untreated or partially treated sewage can introduce heavy metals into rivers, lakes, and oceans [1–5].

Copper is a typical heavy metal found in the environment. It is an essential micronutrient for many organisms but can also biomagnify the food chain. Human activities such as agriculture, mining, and manufacturing are significant sources of copper pollution. Copper-based fertilizers and pesticides used in agriculture; effluents from chemical, pharmaceutical, and paper manufacturing facilities; and corrosion of household plumbing can all contribute to copper pollution [4, 6, 7].

Due to the increasing global production of copper, the concentrations of bioavailable copper compounds have significantly risen in the environment. Soluble forms of copper are the most harmful to human health, and prolonged exposure to elevated levels of copper can lead to severe toxicity, anemia, diabetes, renal disorders, liver damage, and even death. Because of this, it is crucial to use sensitive analytical techniques to monitor copper levels in water resources and effluents to determine if concentrations exceed safe limits and mitigate risks to human populations. The World Health Organization (WHO) has established limits for the maximum acceptable concentration of copper in drinking water to safeguard human health. The World Health Organization (WHO) suggests that the maximum tolerable concentration of copper in drinking water should not exceed 2 mg/L (ppm) as a provisional guideline value. Copper can cause gastrointestinal issues in short-term exposures at levels above 2 ppm. For long-term exposure, the guideline value is set based on a level that does not cause liver damage in sensitive populations [5, 6, 8–10].

Various methods, such as spectrophotometry, voltammetry, sensors, and selective membranes, have been developed to detect copper in water samples. Therefore, this chapter aims to comprehensively assess these technologies for different water samples and evaluate their advantages and disadvantages. **Figure 1** demonstrates a hierarchical graph of these methods.

2. Atomic spectrometric methods

Atomic spectrometric methods (ASMs) use specialized equipment to isolate and measure element-specific signals. These methods can be classified into two categories based on the type of spectral dispersion used: atomic absorption spectroscopy (AAS) and atomic emission spectroscopy (AES). In AAS, ground-state atoms of the analyte element absorb light at specific wavelengths, allowing electronic transitions in the atom's electron orbitals. On the other hand, in AES, thermal energy excites atoms, which then generate emission signals as they return from high-energy states to lower-energy levels. While modern analytical instruments can measure multiple elemental species by inserting equipment set to excite these elements, one

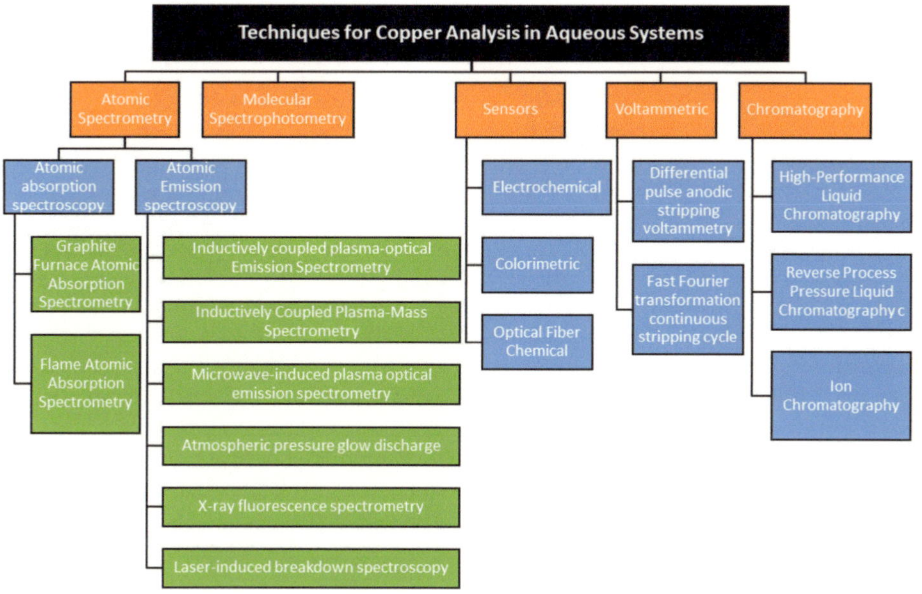

Figure 1.
Techniques for copper analysis in aqueous systems.

significant advantage of atomic emission is the ability to measure numerous elemental species simultaneously [11, 12].

ASM can be divided into two types based on the detection method: photo-spectroscopy and mass-spectroscopy. In photo-spectroscopy, the process is measured by detecting the emitted or absorbed photons. On the other hand, in mass-spectrometric methods, the atom is first ionized to obtain an electric charge and gain electrons. The analytes are then identified and quantified based on their mass-to-charge (m/z) ratio generated from the sample [13].

2.1 Atomic absorption spectroscopy

A widely used technique for analyzing heavy metals is atomic absorption spectrometry (AAS). This encompasses several methods in which free atoms in a gaseous state absorb specific optical radiation, determining the atomic composition. AAS is classified according to the atomization methods into graphite furnace atomic absorption spectrometry (GFAAS), flame atomic absorption spectrometry (FAAS), cold vapor atomic absorption spectrometry (CVAAS), and hydride atomic absorption spectrometry (HAAS). The absorption intensity can be used to evaluate the sample's concentration levels of detected elements. AAS can detect over 70 elements in samples with different physical states, such as solid or solution phases [14–16]. **Figure 2** illustrates the principle of instrumentation for atomic absorption spectrometry. The key features of various atomic absorption spectrometry techniques are presented in **Table 1**.

AAS, FAAS, and GFAAS are standard analytical equipment for detecting copper in liquid samples. Studies worked to improve detection limits, and **Table 2** summarizes the studies conducted on detecting copper (Cu) in water samples using atomic absorption spectroscopy (AAS).

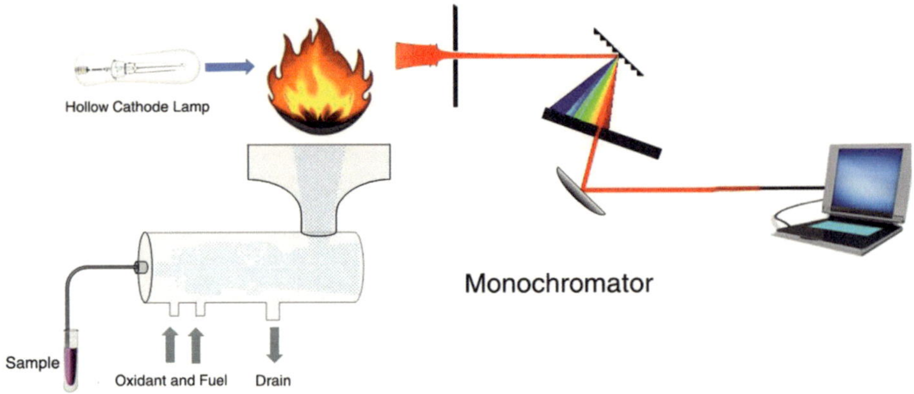

Figure 2.
The principle of instrumentation for atomic absorption spectrometry.

Technique	Principle	Temperature (°C)	Detection limit	Advantages	Disadvantages
Graphite furnace AAS (GFAAS)	Sample heated in graphite furnace	Up to 3000	Pg-ng/mL range	Very low detection limits, minimal sample volumes	Slow analysis, fewer elements detectable
Flame AAS (FAAS)	Sample aspirated into flame	2300 (air-acetylene) 3000 (N2O-acetylene)	Ppb-ppm range	Faster analysis, more elements detectable	Higher detection limits
Cold vapor AAS (CVAAS)	Used for volatile elements like Hg and As	Room temperature	Ppt range	High sensitivity for Hg and As	Only suitable for volatile elements
Hydride AAS (HAAS)	Generates volatile hydrides	Room temperature	Ppb range	Good sensitivity for elements forming hydrides	Only applicable for hydride-forming elements

Table 1.
Comparison of the critical features of various atomic absorption spectrometry techniques.

2.2 Atomic emission spectroscopy

Atomic emission spectroscopy encompasses several techniques that utilize plasma or electrical excitation to generate characteristic atomic emissions for analysis. Major atomic emission spectroscopy techniques include:

- Flame photometry: uses a flame to excite alkali and alkaline earth metals
- Inductively coupled plasma (ICP): uses radio frequency induction to create an argon plasma that excites/ionizes samples. Optical emission spectrometry (ICP-OES) and mass spectrometry (ICP-MS). In ICP-OES, detection relies on emitted photons, whereas in ICP-MS, it depends on the mass-to-charge ratio (m/z). To benefit from the unique features of each detector, many researchers have investigated a combination of OES and MS.

AAS equipment	Sample preparation	Sample type	Detection limit(s)	Reference
GFAAS	Direct determination	Seawater	0.3–0.4 μg/L (single injection) and 0.07 μg/L (multiple injections)	[17]
GFAAS	Dispersive liquid-liquid microextraction	Environmental water	0.01 ng/mL	[18]
FAAS	Direct determination	Wastewater	4 ppb	[19]
FAAS	Solid phase extraction	River water/ sewage water	ng/mL	[9]
FAAS	Dispersive liquid-liquid microextraction	Water samples/ food samples	0.60 μg/L	[20]
FAAS	Direct determination	Pure water spiked with Cu	21 μg/L	[21]
FAAS	Cloud-point preconcentration	Seawater/river water	1.5 μg/L	[22]
FAAS	Cloud-point preconcentration	Seawater/river water	0.04 μg/L	[23]
FAAS	Dispersive liquid-liquid microextraction-slotted quartz tube	Wastewater/tap water/seawater	0.67 μg/L	[24]
FAAS	Cloud-point extraction preconcentration	Water samples	1.5 μg/L	[25]
FAAS	Adsorption onto microcrystalline benzophenone	Water samples	6.9 ng/mL	[26]

Table 2.
Summary of the studies conducted on detecting copper (Cu) in water samples using atomic absorption spectroscopy (AAS).

- Microwave-induced plasma optical emission spectrometry (MIP-OES): uses microwave energy to generate a plasma for excitation

- Atmospheric pressure glow discharge (APGD): operates a plasma at atmospheric pressure and gas temperature to excite sample atoms

- Spark or arc atomic emission spectrometry: utilizes an electric spark or arc for atomic excitation

- X-ray fluorescence (XRF): though not a traditional AES technique, it analyzes characteristic X-ray emissions to detect heavy elements

- Laser-induced breakdown spectroscopy (LIBS): focuses a laser onto the sample to create a plasma and analyzes emitted light

There are two methods for detecting ICP: Optical Emission Spectrometry (ICP-OES) and Mass Spectrometry (ICP-MS). In ICP-OES, detection relies on emitted photons, whereas in ICP-MS, it depends on the mass-to-charge ratio (m/z). To benefit

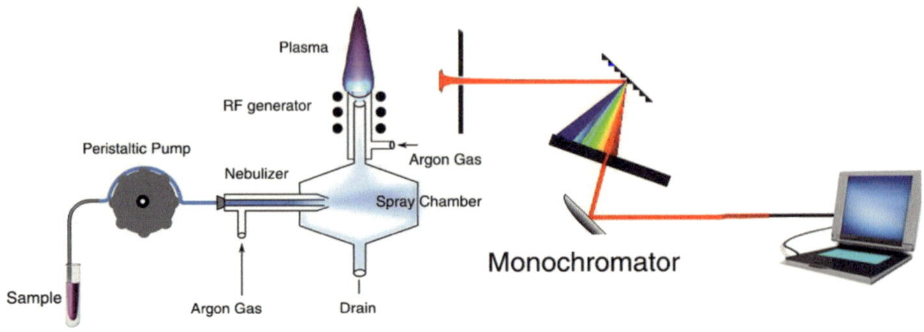

Figure 3.
The principle of instrumentation for inductively coupled plasma.

from the unique features of each detector, many researchers have investigated a combination of OES and MS. This results in more accurate analysis and a broader linear range of calibration curves [27–30]. The principle of instrumentation for inductively coupled plasma is illustrated in **Figure 3**, and a comparison of different atomic emission spectroscopy techniques is detailed in **Table 3**.

3. Molecular spectrophotometric methods

Quantifying copper via colored complex formation with organic reagents offers a simple, rapid, and sensitive detection method for routine analysis. Molecular spectrophotometric methods are widely used for copper detection by measuring the absorption of ultraviolet (UV), visible, or near-infrared (NIR) light by copper-reagent complexes known as chromophores. When copper ions react with specific reagents, characteristic-colored complexes are formed. These chromophores exhibit peak absorbance at specific wavelengths, with the absorbance proportional to the copper concentration as described by Beer's law [12].

Sensitive and selective copper quantification in water samples has been made possible by developing various spectrophotometric reagents [42–44]. Alharthi Al-Saidi [45] synthesized a chromogenic reagent known as 4-amino-3-mercapto-6-[2-(2-thienyl)vinyl]-1,2,4-triazin-5(4H)-one (AMT), which quickly forms a colored complex with copper(II) ions. To prepare a Cu(II) stock solution, Cu(NO3)2 was dissolved in deionized water. Using a HACH DR-6000 spectrophotometer, absorbance spectra of AMT and the Cu-AMT complex were recorded. The AMT-copper(II) complex exhibited a distinct brown color within 10 seconds and was effective over a wide pH range. This method has a detection limit of 0.011 μg/mL, making it suitable for routine copper analysis.

Various spectrophotometric reagents have been studied for detecting copper(II) ions, including azo-Schiff base ligands [46], chloro(phenyl)glyoxime [47], cefixime [48], and 1-(2-pyridylazo)-2-naphthol [49]. These reagents create colored complexes with copper(II), making it possible to detect copper(II) ions at concentrations as low as 0.1 μg/mL. Notably, some of these techniques use non-toxic and environmentally friendly reagents like 5-(4-nitrophenylazo)salicylic acid (NPAS) [50] and polyethyleneimine (PEI) [51]. PEI reacts quickly and selectively with copper(II) ions and exhibits absorbance peaks at 275 and 630 nm. A detection limit of 566 nM has been demonstrated.

DOI: http://dx.doi.org/10.5772/intechopen.1003793

Technique	Principle	Excitation source	Detection	Elements detected	Key features	Limitations	Reference
ICP-OES	Atomic emission spectrometry. Samples are converted to aerosols and introduced into argon plasma, causing excitation and emission of characteristic wavelengths.	Argon plasma torch - radio frequency inductive coil creates a magnetic field that ionizes argon gas into sustained plasma at 6000–10,000 K	Optical emission at characteristic wavelengths measured by spectrometer	Cu, Co, Ni, Mg, Zn	Rapid multi-element analysis; wide linear range; lower detection limits than FAAS	Matrix interferences; spectral interferences	[31]
			Optical emission at characteristic wavelengths measured by spectrometer	Wide range of metals	—	—	[32]
			Optical emission at characteristic wavelengths measured by spectrometer	Cu	Optimized operation conditions to enhance Cu detection limit	—	[33]
ICP-MS	Ionizes sample in plasma, then separates and detects ions based on mass/charge ratio (m/z).	Argon plasma torch - radio frequency inductive coil creates a magnetic field that ionizes argon gas into sustained plasma at 6000–10,000 K	Mass spectrometer separates and detects ions based on m/z	Cu, Cd, Mn, Zn, Pb and isotopes	Very low detection limits; simultaneous analysis of multiple elements; isotope ratio determination	Spectral interferences; sample matrix effects; risk of contamination	[34]
			Mass spectrometer separates and detects ions based on m/z	Heavy metals, including Cu	Preconcentration before analysis improves detection	—	[35]
Microwave-induced plasma optical emission spectrometry (MIP-OES)	Samples are ionized by microwave-induced argon plasma, emitting characteristic optical emissions	Microwave-induced argon plasma	Optical emission spectroscopy	Cd, Cu, Cr, Ni, Pb, Zn	Rapid multi-element analysis; lower detection limits than ICP-OES	Interference from easily ionized elements	[36]

Technique	Principle	Excitation source	Detection	Elements detected	Key features	Limitations	Reference
Atmospheric pressure glow discharge (APGD)	Samples nebulized into aerosol droplets that are introduced into atmospheric pressure glow discharge plasma, causing excitation and characteristic atomic emission	Atmospheric pressure glow discharge argon plasma	Optical emission spectroscopy	Cu	In situ analysis; detection limit of 0.074 μg/L	Only Cu reported; interferences not discussed	[37]
X-ray fluorescence spectrometry	Samples are irradiated with X-rays, causing the emission of secondary fluorescent X-rays that are characteristic of specific elements	X-ray tube excitation source	Detection of emitted fluorescent X-rays	Cu	Rapid quantitative analysis of Cu in water samples	Only Cu reported; interferences not discussed	[38]
					Rapid analysis of Cu content in rocks/ minerals; detection limit 0.51% Cu		[39]
X-ray fluorescence spectroscopy with graphene oxide probes	Samples interact with graphene oxide probes, which enhance characteristic X-ray fluorescence signal of elements	X-ray tube excitation source	Detection of emitted fluorescent X-rays	Cu	Sensitive, quantitative detection of trace Cu; detection limit of 9.8 μg/L	Only Cu was reported; matrix effects were not discussed	[40]
Laser-induced breakdown spectroscopy (LIBS) with chelating resin enrichment	Samples were enriched using chelating resin, then ablated with a laser pulse to form plasma and detect emitted light	Pulsed laser excitation	Optical emission spectroscopy of laser-induced plasma	Pb, Cd, Cu, Cr	Enrichment improves detection limits; detection limits 0.8–2.1 μg/L	Further interferences not discussed	[41]

Table 3.
Comparison of different atomic emission spectroscopy techniques for copper detection in eques samples.

Reagent used	Detection limit	Key features	References
4-amino-3-mercapto-6-[2-(2-thienyl) vinyl]-1,2,4-triazin-5(4H)-one (AMT)	0.011 μg/ml	Rapid color formation, wide pH range	[45]
Azo-Schiff base 1-((4-(1-(2-hydroxyphenyl imino)ethyl)-phenyl)diazenyl) naphthalene-2-ol (HPEDN)	1.7–5.4 μg/ml	Forms colored complex with copper	[46]
Chloro (phenyl) glyoxime	10 μg/L	High molar absorptivity, low detection limit	[47]
Cefixime in 1,4-dioxan-water	0.0319 μg/ml	Rapid analysis, minimal reagents	[48]
1-(2-pyridylazo)-2-naphthol	0.1–2.5 μg/ml	Forms colored chelate complex	[49]
5-(4-nitrophenylazo)salicylic acid and 2,2′-dipyridyl	0.63–5.04 μg/ml	Ternary complex formation	[50]
Polyethyleneimine	566 nM	Fast detection, high sensitivity	[51]

Table 4.
Summary of molecular spectrophotometric methods for the detection of Cu^{2+} in aqueous samples by UV–Vis spectrophotometer.

Advances in chromogenic reagents for copper analyses with improved selectivity are presented in **Table 4**.

4. Sensors

Recent advances in electrochemical, colorimetric, and optical sensor technologies have enabled rapid and sensitive detection of copper in water systems. Electrochemical sensors utilizing square wave anodic stripping voltammetry with electrodes modified by grapefruit peel bio-templates can detect copper at 0.99 μg/L levels in drinking water [52]. Colorimetric sensors employing nanoparticles or organic dyes have achieved detection limits approaching 0.01–6.4 μg/L in aqueous copper solutions and tap water by measuring color changes with UV–vis spectroscopy [53–55]. Optical sensors such as fluorescent Schiff base probes can selectively detect copper at 0.055 μg/L in aqueous samples through fluorescence enhancement upon copper binding [56]. Additionally, surface plasmon resonance biosensors functionalized with chelating agents have reached detection limits around 0.02 μg/L in drinking water [4]. The principles, detection methods, and applications of these sensors provide versatile techniques for sensitive copper quantification in the μg/L range for water quality monitoring. Sensors show immense promise for copper detection across environmental, industrial, and biological systems by combining sensitivity, selectivity, and ease of use. Advances in sensors for copper analyses with improved selectivity are presented in **Table 5**.

5. Electrochemical methods

Electrochemical or (voltammetry) is a reliable method for detecting copper ions in water and wastewater samples. This technique involves applying a time-dependent potential to an electrochemical cell and measuring the resulting current to gain insights into the oxidation or reduction reactions of the subject ions [57].

Recent research has shown that a carbon paste electrode modified with biochar can effectively preconcentrate copper ions on the electrode surface, resulting in a low detection limit, high sensitivity, and stability. This cost-effective approach provides an excellent option for analyzing copper ions [58]. In addition, differential pulse anodic stripping voltammetry and fast Fourier transformation continuous stripping cycle voltammetry are two effective techniques for detecting copper in liquid samples. Differential pulse voltammetry utilizes a gold microelectrode and 0.01 M HNO_3 electrolyte to detect sensitive and fast copper at 1 μM concentrations [59]. The gold microelectrode facilitates the anodic stripping process, where copper is first reduced and deposited on the electrode surface, then oxidized to produce a measurable current peak [60]. In contrast, fast Fourier transform voltammetry employs a carbon paste electrode without an electrolyte. Instead, it uses 2-amino-N-(2-pyridyl methyl)-benzamide as an organic ligand to complex copper ions and transfers them across a supported liquid membrane [61]. The carbon paste electrode enables continuous cyclic stripping measurements for complex copper. While anodic stripping offers very low detection limits, ligand-assisted continuous stripping allows for constant

Sensor type	Principle	Detection method	Sample type	Detection limit	References
Electrochemical sensor using carbon paste electrode prepared from bio-template (grapefruit peels)	Adsorption of copper ions onto carboxyl-functionalized carbon paste electrode	Square wave anodic stripping voltammetry	Drinking water samples	0.99 μg/L	[52]
Colorimetric sensor using silver/dopamine nanoparticles	Color change of silver/dopamine nanoparticles in presence of copper ions	UV–vis absorption spectroscopy	Tap water samples	0.1 μM (equivalent to around 6.4 μg/L)	[53]
Colorimetric probe using gold nanoparticles	Color change caused by aggregation of gold nanoparticles in presence of dopamine	UV–vis spectroscopy	Dopamine solutions	0.5 μM (equivalent to around 0.1 μg/mL)	[54]
Colorimetric sensor using diamine-functionalized SBA-15	Color change caused by coordination of copper ions to diamine groups on SBA-15	UV–vis spectroscopy	Aqueous copper solutions	0.2 μM (equivalent to around 0.01 mg/L)	[55]
Optical sensor using fluorescent Schiff base	Fluorescence enhancement of Schiff base probes upon binding Cu^{2+} and Al^{3+}	Fluorescence spectroscopy	Aqueous solutions	8.7×10^{-9} M for Cu^{2+} (equivalent to 0.055 μg/L)	[56]

Table 5.
Summary of sensors used for the detection of Cu^{2+} in aqueous samples.

Technique	Working electrode	Backing electrolyte/ligand	Stripping	Detection range	Advantages	Limitations	References
Square wave anodic stripping voltammetry	Carbon paste electrode modified with biochar	0.1 mol/L acetate buffer (pH 4.5)	Anodic	0.3–100 μg/L	Low cost, simple preparation, wide linear range	Single use electrode; only tested for Cu	[58]
Differential pulse anodic stripping voltammetry	Gold microelectrode	0.01 M HNO3	Anodic	1 μM Cu(II)	Sensitive, fast response	Interference from other metals	[59, 60]
Fast Fourier transformation continuous stripping cycle voltammetry	Carbon paste electrode	2-Amino-N-(2-Pyridyl Methyl)-Benzamide	Continuous cyclic	—	Can detect complexed Cu, continuous monitoring	Slower than anodic stripping	[61, 62]

Table 6.
Summary of electrochemical methods for copper analyses with improved selectivity.

monitoring and speciation of complex copper [62]. Both techniques provide complementary capabilities for sensitive trace copper quantification and dynamic concentration profiling in water quality monitoring applications. Advances in voltammetric methods for copper analyses with improved selectivity are presented in **Table 6**.

6. Chromatography

Liquid chromatographic (LC) techniques such as high-performance liquid chromatography (HPLC), reverse phase high-pressure liquid chromatography (RP-HPLC), and ion chromatography (IC) can provide sensitive and selective quantitation of copper at trace levels. **Figure 4** illustrates the operational principle of LC.

HPLC with UV detection allows copper detection down to low parts per billion levels [63]. Reverse phase high-pressure liquid chromatography (RP-HPLC) can reach trace detection limits by preconcentrating samples with a tetra (m-aminophenyl) porphyrin (Tm-App) ligand before injection [64–67]. In contrast, ion chromatography uses a mixed mode cation/anion exchange column and ammonium formate/KOH eluent, which retains both hydrated and complex Cu^{2+} species [64, 68]. These chromatographic methods' sensitivity, accuracy, and precision make them ideal for monitoring copper concentrations in environmental waters, evaluating treatment efficiency, and assessing regulatory compliance. When analyzing complex water samples, careful optimization of operating parameters and sample preparation steps are needed to mitigate matrix effects and interferences. Chromatographic techniques are indispensable for reliable copper quantification at environmentally relevant levels in diverse aqueous matrices. Advances in chromatographic techniques for copper analyses with improved selectivity are presented in **Table** 7.

7. Advantages and drawbacks of different detection methods

Various analytical techniques are available for quantitatively detecting copper, each with inherent advantages and limitations. **Table 8** summarizes the advantages and drawbacks of different detection methods for copper analyses.

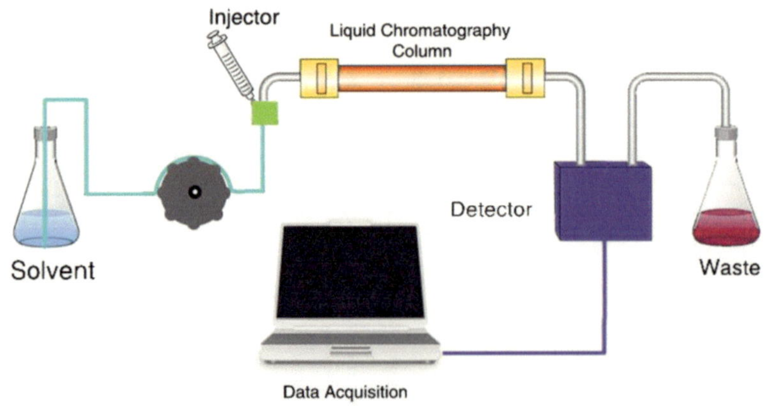

Figure 4.
Operational principle of LC.

Technique	High-performance liquid chromatography (HPLC)	Reverse phase high-pressure liquid chromatography (RP-HPLC)	Ion chromatography (IC)
Stationary phase	C18 reverse-phase column [63]	Nonpolar C18 column [67]	Mixed mode cation/anion exchange column (Ionpac CS5A) [68]
Mobile phase	Buffer with ion pairing agent [63]	Methanol/acetone buffer [67]	Ammonium formate/ KOH eluent [68]
Detection	UV–VIS spectrophotometry [63]	Spectrophotometer at 435 nm [67]	Absorbance at 530 nm after post-column PAR complexation [68]
Preconcentration	Solid phase extraction[63]	Solid phase extraction with Tm-App ligand [66, 67]	—
Copper species	Free copper ions [63]	Hydrated Cu^{2+} [64]	Hydrated and complexed Cu^{2+} [68]
Sensitivity	Low ppb range [63]	Trace levels [64]	~mg/L range [69]
Advantages	High sensitivity, ability to speciate copper [63]	Very sensitive speciation capabilities [64]	Detects complexed copper [70]
Limitations	Matrix effects, interferences [63]	Only detects free hydrated Cu^{2+} [65]	Less sensitive than HPLC [70]

Table 7.
Summary of chromatographic techniques for copper analyses with improved selectivity.

Analytical method	Advantages	Disadvantages	References
Atomic absorption spectrometry (AAS)	• High selectivity • Accurate and reliable quantitation • Fast and simple	• Expensive equipment • Requires skilled staff • Standards needed for calibration • Cannot determine oxidation state or isotope	[14]
Inductively coupled plasma-optical emission spectrometry (ICP-OES)	• Simultaneous multi-element analysis • Fast and simple • Low sample volume	• Expensive equipment • Requires skilled staff • Standards needed for calibration • Cannot determine oxidation state or isotope	[71, 72]
Inductively coupled plasma-mass spectrometry (ICP-MS)	• Very low detection limits • High-resolution capabilities • Simultaneous multi-element analysis • Low sample volume	• Expensive equipment • Requires skilled staff • Standards needed for calibration • Spectral interferences need control	[28, 71, 73]
Microwave-induced plasma optical emission spectrometry (MIP-OES)	• Sensitive detection • Fast analysis • Multi-element capability	• Expensive equipment • Requires skilled personnel • Standards needed for calibration	[36]

Analytical method	Advantages	Disadvantages	References
Atmospheric pressure glow discharge (APGD)	• Cost-effective • Simple operation • Low power requirements	• Limited sensitivity • Difficulty in quantification • May need complementary methods for precise analysis	[37]
X-ray fluorescence spectrometry (XRF)	• Non-destructive • Multi-element analysis • No sample preparation required	• Surface analysis primarily • Limited sensitivity for some elements • Expensive equipment	[38, 39]
X-ray fluorescence spectroscopy with graphene oxide probes	• Enhanced sensitivity • Selective detection • Possibility of real-time analysis	• Requires probe preparation • Limited to surface analysis • Requires skilled personnel	[40]
Laser-induced breakdown spectroscopy (LIBS) with chelating resin enrichment	• Fast, almost real-time analysis • Minimal sample preparation • Multi-element capability • Portable systems available	• Requires skilled personnel • Equipment can be expensive • Standards needed for calibration • Sample enrichment step may add complexity	[41]
Spectrophotometric	• Simple and inexpensive • Precise and sensitive with proper reagents	• Time consuming • Uses large solvent volumes	[45, 46]
Sensors	• Fast real-time analysis • High sensitivity and selectivity • Potential for continuous monitoring	• Sensor lifetime issues • Special calibration solutions • Qualified operators needed	[74]
Voltammetric	• Inexpensive equipment • Easy sensor fabrication • Rapid copper quantitation	• Variability with environmental conditions • Hazardous waste generated • Requires indicator reagents	[59]
Chromatography	• Small sample volumes • High selectivity for complex samples • Environmentally friendly	• Expensive equipment • Requires skilled staff • Standards needed for calibration • Long analysis times • High pressures	[68, 75, 76]

Table 8.
Summary of advantages and drawbacks of different detection methods for copper analyses.

Spectrophotometric and voltammetric methods offer inexpensive and rapid copper analysis but can suffer from interferences and environmental variability. Meanwhile, sophisticated instrumentation like atomic absorption spectrometry (AAS), inductively coupled plasma-optical emission spectrometry (ICP-OES), and inductively coupled plasma-mass spectrometry (ICP-MS) provide sensitive and reliable quantitation, although at substantially higher costs. ICP-MS, in particular,

DOI: http://dx.doi.org/10.5772/intechopen.1003793

enables isotope ratio analysis and limits detection down to parts per trillion levels. Electrochemical sensors present a promising on-site or continuous monitoring approach given their real-time measurement capabilities. However, widespread implementation still needs to address sensor robustness and operator expertise. Where sample complexity is high, chromatographic separation before element-specific detection can improve selectivity and sensitivity without extensive sample pretreatment. Critical parameters such as target detection limits, available resources, and sample matrix effects are needed to select the optimal copper detection methodology for a given application.

Abbreviations

AAS	atomic absorption spectroscopy
AES	atomic emission spectroscopy
FAAS	flame atomic absorption spectrometry
GFAAS	graphite furnace atomic absorption spectrometry
CVAAS	cold vapor atomic absorption spectrometry
HAAS	hydride atomic absorption spectrometry
ICP-OES	inductively coupled plasma-optical emission spectrometry
ICP-MS	inductively coupled plasma-mass spectrometry
MIP-OES	microwave-induced plasma optical emission spectrometry
APGD	atmospheric pressure glow discharge
XRF	X-ray fluorescence spectroscopy
LIBS	laser-induced breakdown spectroscopy
UV-Vis	ultraviolet-visible spectroscopy
HPLC	high-performance liquid chromatography
RP-HPLC	reverse phase high-pressure liquid chromatography
IC	ion chromatography

Author details

Ahmed Elkhatat
Department of Chemical Engineering, College of Engineering, Qatar University, Doha, Qatar

*Address all correspondence to: ahmed.elkhatat@qu.edu.qa

References

[1] Kinuthia GK et al. Levels of heavy metals in wastewater and soil samples from open drainage channels in Nairobi, Kenya: Community health implication. Scientific Reports. 2020;**10**(1):1-13

[2] Sobhanardakani S, Tayebi L, Farmany A. Toxic metal (Pb, Hg and As) contamination of muscle, gill and liver tissues of Otolithes rubber, Pampus argenteus, Parastromateus niger, Scomberomorus commerson and Onchorynchus mykiss. World Applied Sciences Journal. 2011;**14**(10):1453-1456

[3] Wołowiec M et al. Removal of heavy metals and metalloids from water using drinking water treatment residuals as adsorbents: A review. Minerals. 2019;**9**(8):487

[4] Pesavento M et al. An optical fiber chemical sensor for the detection of copper (II) in drinking water. Sensors. 2019;**19**(23):5246

[5] WHO. Guidelines for Drinking-Water Quality. 4th ed. Switzerland: W.H. Organization; 2011. pp. 1-564

[6] Government of Western Australia, D.o.P.H.G.-D. Copper in drinking water 2021. Available from: https://ww2.health.wa.gov.au/Articles/A_E/Copper-in-drinking-water [Accessed: May 14, 2016]

[7] Briffa J, Sinagra E, Blundell R. Heavy metal pollution in the environment and their toxicological effects on humans. Heliyon. 2020;**6**(9):e04691

[8] da Costa Lopes AM, Bogel-Łukasik R. Acidic ionic liquids as sustainable approach of cellulose and lignocellulosic biomass conversion without additional catalysts. ChemSusChem. 2015;**8**(6):947-965

[9] Taylor AA et al. Critical review of exposure and effects: Implications for setting regulatory health criteria for ingested copper. Environmental Management. 2020;**65**(1):131-159

[10] Malik LA et al. Detection and removal of heavy metal ions: A review. Environmental Chemistry Letters. 2019;**17**(4):1495-1521

[11] Broekaert JAC. Spectrometric instrumentation. In: Analytical Atomic Spectrometry with Flames and Plasmas. New York City, United States: Wiley & Sons, Ltd; 2006. pp. 34-93

[12] Bachmann LM, Miller WG. Spectrophotometry. In: Clarke W, Marzinke MA, editors. Contemporary Practice in Clinical Chemistry. Cambridge, Massachusettsm United States: Academic Press; 2020. pp. 119-133

[13] Thomas SN. Mass spectrometry. In: Clarke W, Marzinke MA, editors. Contemporary Practice in Clinical Chemistry. Cambridge, Massachusettsm United States: Academic Press; 2019. pp. 171-185

[14] Zhang Y et al. Technologies for detection of HRPs in wastewater. In: High-Risk Pollutants in Wastewater. Cambridge, Massachusettsm United States: Elsevier; 2020. pp. 79-100

[15] Baysal A, Ozbek N, Akman S. Determination of trace metals in waste water and their removal processes. In: Waste Water-Treatment Technologies and Recent Analytical Developments. London, UK: InTech; 2013. pp. 145-171

[16] Michalke B, Nischwitz V. Liquid chromatography: Chapter 22. In: Speciation and Element-Specific

Detection. Cambridge, Massachusettsm United States: Elsevier Inc; 2013

[17] Huang S-D, Shih K-Y. Direct determination of copper in sea-water with a graphite furnace atomic absorption spectrometer. Spectrochimica Acta Part B: Atomic Spectroscopy. 1993;**48**(12):1451-1460

[18] Han Q et al. Determination of ultra-trace amounts of copper in environmental water samples by dispersive liquid-liquid microextraction combined with graphite furnace atomic absorption spectrometry. Separations. 2023;**10**(2):93

[19] Khayatian G, Moradi M, Hassanpoor S. MnO2/3MgO nanocomposite for preconcentration and determination of trace copper and lead in food and water by flame atomic absorption spectrometry. Journal of Analytical Chemistry. 2018;**73**(5):470-478

[20] Bagherian G et al. Determination of copper (II) by flame atomic absorption spectrometry after its perconcentration by a highly selective and environmentally friendly dispersive liquid–liquid microextraction technique. Journal of Analytical Science and Technology. 2019;**10**(1):1-11

[21] Lima RT et al. Determination of copper at wide range concentrations using instrumental features of high-resolution continuum source flame atomic absorption spectrometry. Eclética Química. 2010;**35**(4):87-92

[22] Lemos VA et al. Determination of copper in water samples by atomic absorption spectrometry after cloud point extraction. Microchimica Acta. 2007;**157**(3-4):215-222

[23] Goudarzi N. Determination of trace amounts of copper in river and sea water samples by flame atomic absorption spectrometry (FAAS) after cloud-point preconcentration. Journal of the Brazilian Chemical Society. 2007;**18**(7):1348-1352

[24] Özzeybek G et al. Sensitive determination of copper in water samples using dispersive liquid-liquid microextraction-slotted quartz tube-flame atomic absorption spectrometry. Microchemical Journal. 2017;**132**:406-410

[25] Snigur DV, Dubovyi VP, Chebotarev AN. Atomic-absorption determination of copper(II) in water samples after its cloud-point extraction preconcentration. Moscow University Chemistry Bulletin. 2021;**75**(6):328-332

[26] Kim B, Choi H. Determination of copper by atomic absorption spectrophotometry after solid-liquid separation by adsorption of its 1-Nitroso-2-naphthol complex onto microcrystalline benzophenone. Analytical Letters. 1999;**32**(5):995-1009

[27] Deng Y et al. Sharing one ICP source for simultaneous elemental analysis by ICP-MS/OES: Some unique instrumental capabilities. Microchemical Journal. 2017;**132**:401-405

[28] Poirier L et al. Application of ICP-MS and ICP-OES on the determination of nickel, vanadium, iron, and calcium in petroleum crude oils via direct dilution. Energy & Fuels. 2016;**30**(5):3783-3790

[29] Garbarino JR, Taylor HE, Batie WC. Simultaneous determination of major and trace elements by inductively coupled plasma-mass spectrometry/optical emission spectrometry. Analytical Chemistry. 1989;**61**(7):793-796

[30] Sawatari H et al. Versatile simultaneous multielement measurement system with combination of ICP-MS and

ICP-AES through optical fiber. Bulletin of the Chemical Society of Japan. 1995;**68**(6):1635-1640

[31] Ranjbar L et al. Ionic liquid based dispersive liquid-liquid microextraction combined with ICP-OES for the determination of trace quantities of cobalt, copper, manganese, nickel and zinc in environmental water samples. Microchimica Acta. 2012;**177**(1-2):119-127

[32] Giersz J, Bartosiak M, Jankowski K. Sensitive determination of Hg together with Mn, Fe, Cu by combined photochemical vapor generation and pneumatic nebulization in the programmable temperature spray chamber and inductively coupled plasma optical emission spectrometry. Talanta. 2017;**167**:279-285

[33] Dimpe K et al. Evaluation of sample preparation methods for the detection of total metal content using inductively coupled plasma optical emission spectrometry (ICP-OES) in wastewater and sludge. Physics and Chemistry of the Earth, Parts A/B/C. 2014;**76**:42-48

[34] Venkatesan AK et al. Using single-particle ICP-MS for monitoring metal-containing particles in tap water. Environmental Science: Water Research & Technology. 2018;**4**(12):1923-1932

[35] Xing G et al. Analysis of trace metals in water samples using NOBIAS chelate resins by HPLC and ICP-MS. Talanta. 2019;**204**:50-56

[36] Baraud F et al. New approach for determination of Cd, Cu, Cr, Ni, Pb, and Zn in sewage sludges, fired brick, and sediments using two analytical methods by microwave-induced plasma optical spectrometry and induced coupled plasma optical spectrometry. SN Applied Sciences. 2020;**2**(9):1536

[37] Yang H et al. In situ detection of trace heavy metal Cu in water by atomic emission spectrometry of nebulized discharge plasma at atmospheric pressure. Applied Sciences. 2022;**12**(10):4939

[38] Hua Z. Rapid determination of copper content in condenser cooling water by X-ray fluorescence spectrometer. In: China International Conference on Electricity Distribution (CICED). Shanghai, China: IEEE; 2021. pp. 235-237

[39] Escárate P et al. X-ray fluorescence spectroscopy for accurate copper estimation. Minerals Engineering. 2015;**71**:13-15

[40] Liu J-H et al. Quantitative detection of trace copper by using graphene oxide and X-ray fluorescence spectroscopy. Nano. 2021;**16**(06):2150066

[41] Wang J et al. Highly sensitive detection of heavy metal elements using laser-induced breakdown spectroscopy coupled with chelating resin enrichment. Chemosensors. 2023;**11**(4):228

[42] Satheesh K, Rao VS. A study on spectrophotometric determination of copper from wastewater and its removal using magnatite nanoparticles. Archives of Applied Science Research. 2016;**8**(8):31-36

[43] Fu D, Yuan D. Spectrophotometric determination of trace copper in water samples with thiomichlersketone. Spectrochimica Acta Part A: Molecular and Biomolecular Spectroscopy. 2007;**66**(2):434-437

[44] Awual MR, Hasan MM. Colorimetric detection and removal of copper (II) ions from wastewater samples using tailor-made composite adsorbent. Sensors and Actuators B: Chemical. 2015;**206**:692-700

[45] Alharthi S, Al-Saidi H. Spectrophotometric determination of trace concentrations of copper in waters using the chromogenic reagent 4-amino-3-mercapto-6-[2-(2-thienyl) vinyl]-1, 2, 4-triazin-5 (4H)-one: Synthesis, characterization, and analytical applications. Applied Sciences. 2020;**10**(11):3895

[46] Raafid E, Al-Da'amy MA, Kadhim SH. Spectrophotometric determination of Cu(II) in analytical sample using a new chromogenic reagent (HPEDN). Indonesian Journal of Chemistry. 2020;**20**(5):1080

[47] Turkoglu O, Soylak M. Spectrophotometric determination of copper in natural waters and pharmaceutical samples with chloro (phenyl) glyoxime. Journal of the Chinese Chemical Society. 2005;**52**(3):575-579

[48] Sharma S et al. UV spectrophotometric determination of Cu (II) in synthetic mixture and water samples. Journal of the Chinese Chemical Society. 2010;**57**(4A):622-631

[49] Sarker KC. Determination of trace amount of copper (Cu) using UV-VIS spectrophotometric method. International Journal of Chemical Studies. 2013;**1**(1):5-15

[50] Hashem E, Seleim M, El-Zohry AM. Environmental method for spectrophotometric determination of copper (II). Green Chemistry Letters and Reviews. 2011;**4**(3):241-248

[51] Wen T et al. A facile, sensitive, and rapid spectrophotometric method for copper (II) ion detection in aqueous media using polyethyleneimine. Arabian Journal of Chemistry. 2017;**10**:S1680-S1685

[52] Fattahi P et al. A review of organic and inorganic biomaterials for neural interfaces. Advanced Materials. 2014;**26**(12):1846-1885

[53] Ma Y-r, Niu H-y, Cai Y-q. Colorimetric detection of copper ions in tap water during the synthesis of silver/dopamine nanoparticles. Chemical Communications. 2011;**47**(47):12643-12645

[54] Chen Z et al. Gold nanoparticle based colorimetric probe for dopamine detection based on the interaction between dopamine and melamine. Microchimica Acta. 2015;**182**(5-6):1003-1008

[55] Wang Z et al. Colorimetric detection of copper and efficient removal of heavy metal ions from water by diamine-functionalized SBA-15. Dalton Transactions. 2014;**43**(22):8461-8468

[56] Sharma S et al. Method for highly selective, ultrasensitive fluorimetric detection of Cu2+ and Al3+ by Schiff bases containing o-phenylenediamine and o-aminophenol. Methods. 2023;**217**:27-35

[57] Elkhatat AM et al. Recent trends of copper detection in water samples. Bulletin of the National Research Centre. 2021;**45**:218

[58] Oliveira PR et al. Electrochemical determination of copper ions in spirit drinks using carbon paste electrode modified with biochar. Food Chemistry. 2015;**171**:426-431

[59] Zhuang J et al. Determination of trace copper in water samples by anodic stripping voltammetry at gold microelectrode. International Journal of Electrochemical Science. 2011;**6**:4690-4699

[60] Javanbakht M et al. Determination of picomolar silver concentrations by differential pulse anodic stripping voltammetry at a carbon paste electrode modified with phenylthiourea-functionalized high ordered nanoporous silica gel. Electrochimica Acta. 2009;**54**(23):5381-5386

[61] Mofidi Z et al. Determination of diclofenac using electromembrane extraction coupled with stripping FFT continuous cyclic voltammetry. Analytica Chimica Acta. 2017;**972**:38-45

[62] Romero-Cano LA et al. Electrochemical detection of copper in water using carbon paste electrodes prepared from bio-template (grapefruit peels) functionalized with carboxyl groups. Journal of Electroanalytical Chemistry. 2019;**837**:22-29

[63] Yuan C-g et al. Speciation of heavy metals in marine sediments from the East China Sea by ICP-MS with sequential extraction. Environment International. 2004;**30**(6):769-783

[64] Yang Y, Guangyu Y, Lin Q. Determination of heavy metal ions in Chinese herbal medicine by microwave digestion and RP-HPLC with UV-vis detection. Microchimica Acta. 2004;**144**(4):297-302

[65] Rekhi H et al. A review on recent applications of high-performance liquid chromatography in metal determination and speciation analysis. Critical Reviews in Analytical Chemistry. 2017;**47**(6):524-537

[66] Hu Q et al. Determination of copper, nickel, cobalt, silver, lead, cadmium, and mercury ions in water by solid-phase extraction and the RP-HPLC with UV-vis detection. Analytical and Bioanalytical Chemistry. 2003;**375**(6):831-835

[67] Daud JM, Alakili IM. High performance liquid chromatographic separations of metal pyrrolidine dithiocarbamate complexes. Malaysian Journal of Analytical Sciences. 2001;**7**(1):113-120

[68] Kulisa K, Polkowska-Motrenko I, Dybczynski R. Determination of transition metals by ion chromatography. In: Radiochemistry, Stable Isotopes, Nuclear Analytical Methods, Chemistry in General. Warszawa: Institute of Nuclear Chemistry and Technology; 1999. pp. 74-76

[69] Haddad PR, Jackson PE. Ion Chromatography: Principles and Applications. Amsterdam, The Netherlands: Elsevier; 1990

[70] Lu H et al. On-line pretreatment and determination of Pb, Cu and Cd at the μg l– 1 level in drinking water by chelation ion chromatography. Journal of Chromatography A. 1998;**800**(2):247-255

[71] Balcaen L et al. Inductively coupled plasma–tandem mass spectrometry (ICP-MS/MS): A powerful and universal tool for the interference-free determination of (ultra) trace elements–A tutorial review. Analytica Chimica Acta. 2015;**894**:7-19

[72] National Research Council. Forensic Analysis: Weighing Bullet Lead Evidence. Washington, DC: National Academies Press; 2004

[73] Wilschefski SC, Baxter MR. Inductively coupled plasma mass spectrometry: Introduction to analytical aspects. The Clinical Biochemist Reviews. 2019;**40**(3):115

[74] Hung S-C, Lu C-C, Wu Y-T. An investigation on design and characterization of a highly selective LED

optical sensor for copper ions in aqueous solutions. Sensors. 2021;**21**(4):1099

[75] Kaiser RE. Gas chromatography in environmental analysis. Journal of Chromatographic Science. 1974;**12**(1):36-39

[76] Jackson PE. Ion chromatography in environmental analysis. In: Encyclopedia of Analytical Chemistry: Applications, Theory and Instrumentation. New York City, United States: Wiley & Sons, Ltd.; 2006

Chapter 9

Cu Recovery from E-wastes

Hadi Sharifidarabad

Abstract

Due to technological development and increased production efficiency in all industries, recovery of metals from secondary sources is one of the most important issues. Copper is used in a variety of residential and industrial applications, including power generation and transmission (infrastructure), building wiring, transportation, industrial machinery, commercial durables, and electrical and electronic products due to its unique physical and chemical properties, such as high ductility, malleability, electrical and thermal conductivity, and excellent corrosion resistance. For these reasons, electronic waste is a well-known secondary resource rich in copper. This topic focuses on the statistical study of electronic waste and the metals it contains, the mineralogical and elemental identification of copper in electronic waste, and the study of the steps and methods for recovering copper from electronic waste, especially pyrometallurgy, hydrometallurgy, biohydrometallurgy, and their combination.

Keywords: recycling, E-wastes, copper, pyrometallurgy, hydrometallurgy, biohydrometallurgy

1. Introduction

E-waste, technically known as waste electrical and electronic equipment (WEEE), includes non-functioning and discarded electrical or electronic equipment and its discarded components. It is a significant global problem whose increasing generation rates are being closely monitored by the scientific research community [1, 2].

E-waste is classified on the basis of functional similarity, material composition, average weight, and obsolescence characteristics. In the 2002 European Union Directive, e-waste was divided into ten classes, including toys, medical equipment, sports equipment, and vending machines. These classifications help to better understand and manage the different types of e-waste [1].

In 2016, Baldé et al. reported about 48.2 million tons (6.1 kg per capita per year) of e-waste, with the largest contribution coming from Asia (about 41%), followed by the United States (around 29%) and the European Union (EU) (about 27%) [3, 4]. According to the Global E-waste Monitor 2020, around 53.6 million tons of e-waste were generated worldwide in 2019, which corresponds to 7.3 kilograms per person. It is predicted that this amount will exceed 74 million tons by 2030, with an estimated annual increase of 2 million tons. Europe is the leader in the generation of e-waste per person (16.2 kilograms per person), while Asia recorded the highest generation of e-waste, with around 24.9 million tons in 2019. China, India, Japan, and Indonesia are the main producers of e-waste generation in Asia [1, 5]. **Figure 1** illustrates the global

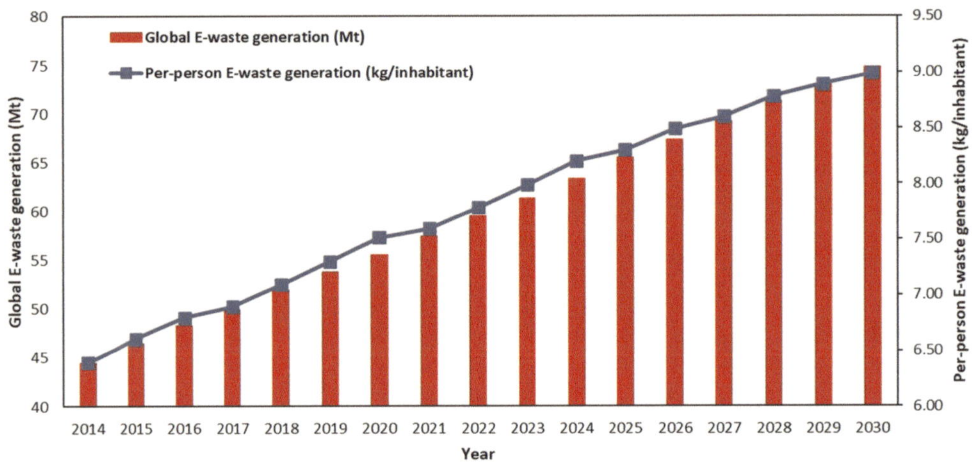

Figure 1.
Global E-waste generation from 2014 to 2030 [6].

generation of e-waste and the forecast scenarios for the generation of e-waste from 2014 to 2030 [7]. The increasing generation of e-waste can be attributed to factors such as higher production and consumption rates, industrialization and urbanization, availability of cheaper versions, shorter product lifespan, and technological advances [5]. In the United States, China, and European countries, the typical lifespan of electronic devices such as smartphones is less than two years [1]. The increase in e-waste production can be attributed to technological advances, industrial automation, progress in information and communication technology (ICT), economic growth rates, and competition for the affordability of electrical and electronic equipment [8, 9]. In addition, this generation pattern is closely linked to income levels, social status, and geographical location, as the growth rate of e-waste is also influenced by advanced materials and manufacturing technologies, rapid market penetration, and a stable economy [7].

Electronic and electrical equipment (EEE) covers a wide range of items, from essential to luxury, and is used in both households and public institutions.

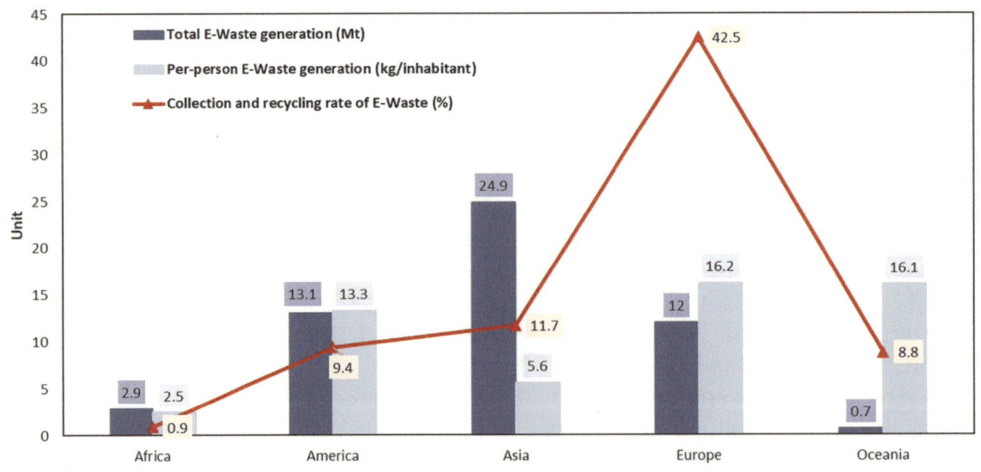

Figure 2.
Generation and recycling rate of E-waste according to the continent [7].

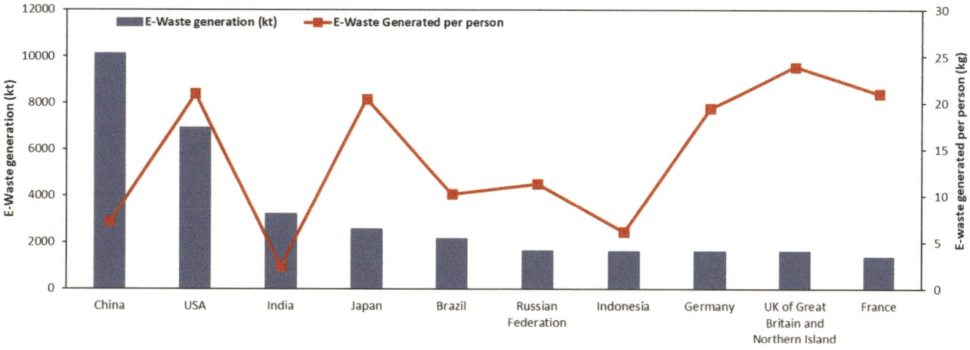

Figure 3.
Top ten E-waste producers in 2019 [2, 5].

Figure 2 shows that while Europe, Oceania, and the Americas have the highest per capita generation of e-waste, the total amount of e-waste produced there is relatively low. This discrepancy can be attributed to factors such as income levels, the availability of recycling and recovery facilities, and the cross-border movement of e-waste from industrialized to developing countries [7]. **Figure 3** shows the largest e-waste producers in 2019.

The different classes of waste, electrical and electronic equipment, contribute to the total volume of e-waste. Small appliances, large appliances, and heat exchangers contribute the most. The rapid progress in the smartphone industry has meant that the lifespan of electronic devices is getting shorter and shorter, making them a major contributor to e-waste generation [1].

Of the total e-waste generated (53.6 million tons), only 9.3 million tons (17.4%) were formally recycled, while the remaining 44.3 million tons (82.6%) remained untreated. Although the recycling rate has improved compared to 2014 (1.8 million tons), it is not enough to offset the annual growth rate of 2 million tons in the generation of e-waste. Europe had the highest recycling rate at 42.5%, followed by Asia (11.7%), America (9.4%), Oceania (8.8%), and Africa (0.9%). Obviously, the stricter regulations for the management of e-waste in industrialized countries contribute to higher recycling rates [1]. Currently, the treatment of printed circuit boards (WPCBs), which account for about 3–6% of total e-waste, poses a major environmental challenge [10].

Tackling the problem of e-waste requires a comprehensive approach involving governments, manufacturers, consumers, and the recycling industry. Initiatives such as extended producer responsibility (EPR), proper collection and disposal schemes, and public awareness campaigns play an important role in promoting the responsible management of e-waste. By adopting sustainable practices and embracing the circular economy principles, we can reduce the environmental and health impacts associated with e-waste and create a more sustainable future.

2. Chemical composition of e-waste

E-waste, also known as electronic waste, has been extensively studied by researchers who have categorized it into different components and materials. These categories include iron (ferrous), non-ferrous metals and compounds, plastics, and other materials such as rare earth elements (REEs). Iron and steel make up around 50% of the weight of e-waste, while plastics such as polycarbonates or polystyrene make up around 10–30%. E-waste also contains a variety of metals, including copper, zinc, nickel, lead, arsenic, mercury, chromium, cadmium, gold, silver, and palladium. Interestingly, the amount of copper found in e-waste can be up to 40 times greater than the amount found in natural copper ores [1].

In recent decades, the recycling of waste electrical and electronic equipment (WEEE) and the identification of innovative business models that can bring significant benefits have become increasingly important [11]. Within WEEE, printed circuit boards (PCBs) account for about 3% by weight and contain about 60 different elements [12]. Among the various components found in WPCBs, metals account for about 30% of their weight, which is significantly more than the metal content in natural minerals. These metals include base metals such as copper (Cu) and zinc (Zn), precious metals such as gold (Au), silver (Ag), and palladium (Pd) as well as heavy metals [13]. **Figure 4** illustrates common materials that can be recovered from e-waste.

The potential of e-waste as a source of valuable and critical elements such as iron, aluminum, copper, gold, silver, platinum group metals (PGMs), and various plastics has been recognized. The raw materials in e-waste were estimated at USD 57 billion in 2019, assuming an ideal collection and recycling rate of 100%. However, only 17.4% of e-waste was collected and recycled that year, with the majority ending up in landfill, landfill sites, or incinerated [2].

Copper is one of the most important metals in e-waste. E-waste is reported to be the largest source of copper-containing waste worldwide, followed by construction and demolition waste, electrical industry waste, and end-of-life vehicles. Different countries have different classification schemes for e-waste. In Europe, for example, the EU's WEEE Directive divides e-waste into six categories according to type and size, while in Japan, e-waste is categorized by size and includes both large and small household appliances. South Korea categorizes e-waste by weight [2]. WPCBs come from various sources, including PCB manufacturers, electrical and electronic equipment (EEE) manufacturers, and e-waste recycling facilities of different sizes. The composition of PCBs is typically 30% ceramics, 40% metals, and 30% organics. The metal content in PCBs varies depending on the type. For example, desktop PCBs usually contain 20% copper and 250 parts per million (ppm) gold, 210 ppm palladium, and 13% copper. TV circuit boards, on the other hand, contain 10% copper, 20 ppm

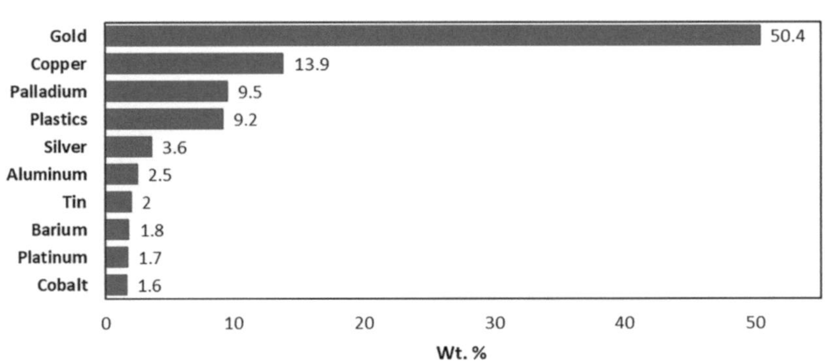

Figure 4.
Common materials that can be recovered from e-waste [11].

gold, and 110 ppm palladium [10]. It is obvious that the different types of e-waste contain different amounts of copper. The copper content in PCBs of cell phones, PC scrap, LCD notebooks, tablets, TV scrap, and calculator scrap is 39.6% [14], 2.84% [15], 6.93% [16], 135 g/unit [11], 27 g/unit [11], 3.8% [16], and 3% [16].

3. Why we should recycle e-waste

The expansion of technology and the increasing affinity to purchase multiple electronic products such as smartphones, tablets, laptops, and game consoles has led to a rapid increase in global consumption, estimated at around 2 million tons annually [5]. This increase in electronic waste (e-waste) poses a significant threat to local communities as it contains toxic metallic and non-metallic components that can have negative effects on human health. These negative effects include DNA damage and mutations, cardiovascular problems, skin diseases, cancer, hearing problems, neurological disorders, learning difficulties, and respiratory effects [1].

Addressing the aggravation of e-waste is a major concern to ensure human health and achieve ecosystem sustainability. However, a more optimistic view of e-waste is to consider it as an additional resource for metals, which are essentially non-natural ores. An example of this approach is the Tokyo 2020 medal project in Japan, which collected around 78,985 tons of e-waste across the country. Around 32 kg of gold, 3500 kg of silver, and 2200 kg of bronze were recovered from this project for the production of Olympic and Paralympic medals [17].

In India, the third largest e-waste producer in the world with 3.2 million tons and limited mineral resources, proper recycling of precious metal components should be a top priority as it enables effective waste management and can potentially improve the country's economy. Currently, only about 20% of e-waste is recycled in India, with 178 registered e-waste recyclers. Earlier, recycling processes were largely informal and involved manual dismantling and open burning, resulting in low efficiency in metal recovery and significant pollution to the ecosystem [18].

Copper is an important non-ferrous metal with unique physical and chemical properties such as good ductility, high thermal and electrical conductivity, and corrosion resistance, and it is widely used in various industries. However, despite the relatively large copper resources in China, the overall copper resources, especially the high-grade copper resources, are still scarce. Therefore, the comprehensive recycling of copper-containing PCB waste is of great social and economic significance for environmental protection and resource recycling [19].

Improper disposal of e-waste can cause significant damage to the environment and human health. In addition, the processing cost of natural ores is much higher (about 10 to 160 times) than the cost of recycling. Due to these factors, the e-waste recycling market has experienced tremendous growth and was valued at USD 41.97 billion in 2019, with projections that it will reach USD 102.62 billion by 2027, growing at a compound annual growth rate (CAGR) of 11.9% from 2020 to 2027. As a result, e-waste recycling has become an emerging business globally [1].

Given the very heterogeneous composition of e-waste, there is no universal treatment technique for its recycling or metal recovery. Furthermore, e-waste contains multiple metallic components, which requires a comprehensive understanding of the process dynamics and interactive mechanisms to maximize recovery potential. Temperature plays a crucial role in these processes, and knowledge of the process thermodynamics of base metals and analysis of the composition of the intermediates formed are essential for process optimization [20]. Future research efforts should focus on these important areas in order to provide better methods for the recovery of metals from e-waste.

4. The way of recovery

A systematic and formal approach to e-waste recycling involves three main processes: e-waste collection, pre-treatment, and metal recovery. **Figure 5** illustrates these processes. In contrast, informal recycling methods are characterized by unscientific and unethical practices, often lacking safety measures [21]. These informal processes not only pollute natural resources but also pose a risk to human health. For example, the informal incineration method of recovering economically valuable metals such as copper (Cu), aluminum (Al), gold (Au), lead (Pb), mercury (Hg), cadmium (Cd), palladium (Pd), and platinum (Pt) leads to the release of hydrocarbons that pollute the air. Cu in particular is highly toxic and dangerous to the ecosystem, as recognized in the RoHS (Restriction of certain Hazardous Substances) Directive. Open burning, the preferred method for recovering copper in informal recycling, leads to the release of hazardous compounds into the environment [22]. Highly toxic chemicals such as hydrochloric acid (HCl), nitric acid (HNO3), sulfuric acid (H_2SO_4), and cyanide are also frequently used in the recycling of metals such as silver (Ag), gold (Au), and palladium (Pd). The vapors produced during these processes are highly polluting to the atmosphere [23]. In addition, the efficiency of metal extraction in informal processes is less than 80%, whereas recovery rates of over 95% can be achieved in the formal processing of e-waste. During pre-treatment, the e-waste is physically broken down and then desoldered, shredded, and fractionated into metallic and non-metallic components. The metallic fractions are subjected to further recycling processes, which may include pyrometallurgical, hydrometallurgical, and biohydrometallurgical methods [1].

Various methods have been used to recycle printed circuit boards, including mechanical, pyrometallurgical, hydrometallurgical, and biological processes (**Figure 6**). Physical recovery processes effectively separate metals from non-metals; they do not produce pure metal products and must be combined with hydrometallurgical processes. Furthermore, physical recovery is an important step in the pre-treatment of electronic waste recycling [1, 19]. In a typical pyrometallurgical process, e-waste is melted in high-temperature furnaces, and the resulting slag, which contains refractory materials, is processed to recover precious or base metals [24]. In hydrometallurgical processes, special activating agents such as halides, thiourea, thiosulphate, cyanide, and acids are used to leach metals by forming stable metal complexes [25]. In biohydrometallurgy, selective microorganisms are used either in the presence of cells or in cell-free extraction media to extract metals. This microbially driven process is efficient, environmentally friendly, and cost-effective. It requires

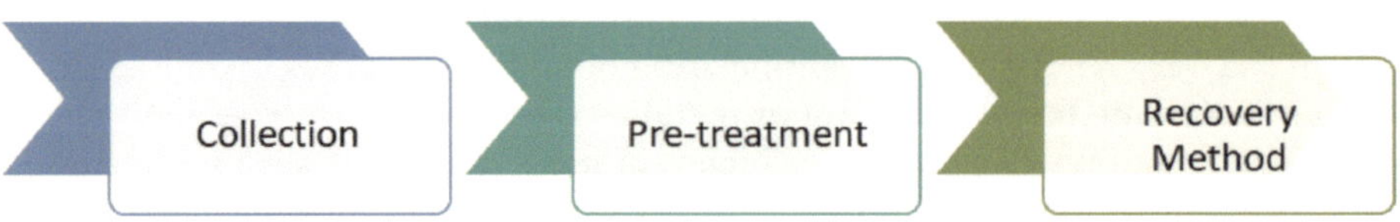

Figure 5.
Main process of e-waste recycling formal approach.

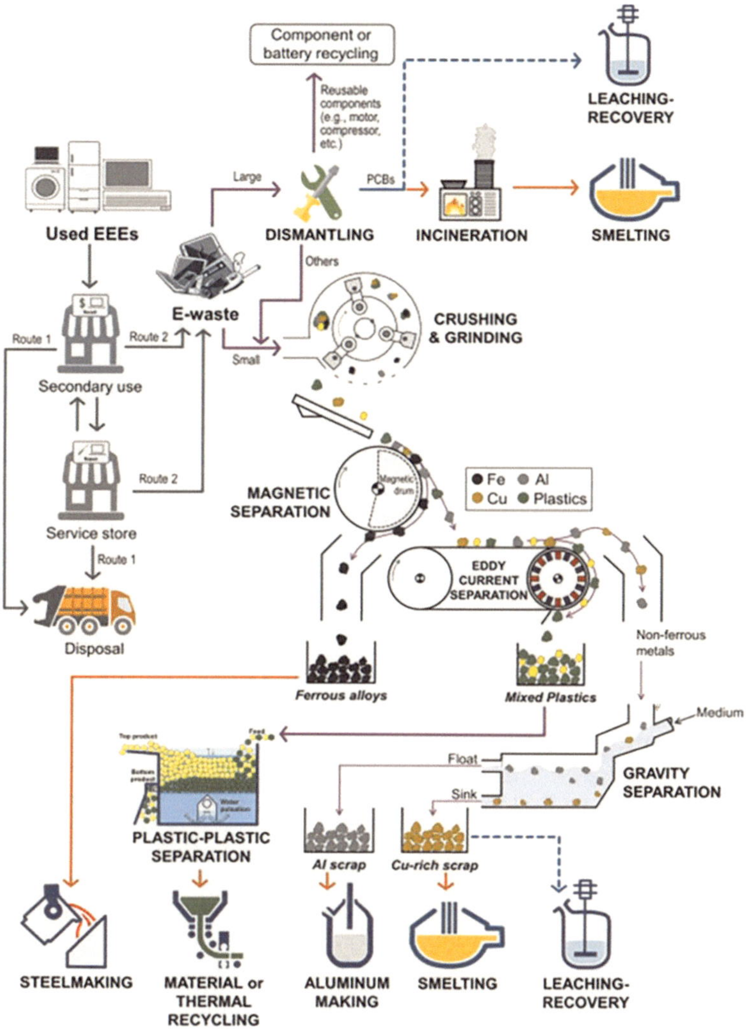

Figure 6.
Schematic flowchart of e-waste recycling methods [2].

minimal labor or energy and can be carried out with minimal industrial equipment, even at low metal concentrations [26]. When considering formal recycling methods, it is important to critically analyze their applicability and potential drawbacks in order to select the optimal method(s) for e-waste management [1].

4.1 Collection

The collection of e-waste or waste electrical and electronic equipment (WEEE) poses a major challenge for e-waste recycling. In contrast to other types of waste, the collection and sorting of e-waste requires considerable effort due to the wide variety of materials used in compact form in electrical and electronic equipment [27]. Collection is the first stage of the e-waste recycling process and is a challenge in both low- and high-income countries. In 2017, France and Italy had e-waste collection rates of 43.2% and 21.7%, respectively, while China, South Africa, and Chile had lower rates of 17.9%, 5.6%, and 0.4%, respectively [28]. The collection of e-waste is particularly difficult in low- and middle-income countries because there is no national legislation or policy specifically for e-waste, and the infrastructure for waste management is lacking or underdeveloped. According to [5], only 78 countries worldwide have implemented national e-waste management policies, laws, or regulations, and the informal sector plays an important role in the management of e-waste. Consequently, approximately 82.6% of e-waste worldwide remains undocumented, with Africa, Oceania, South and North America, Asia, and Europe, having rates of 99.1%, 91.2%, 90.6%, 88.3%, and 57.5%, respectively [5]. Governments can enhance e-waste collection and recycling rates by enacting laws that penalize the disposal of e-waste with household waste. In Iran, the responsibility for the disposal of industrial waste lies with the manufacturers, and the providers of such services must also take on aspects of waste disposal for their products. In Japan, for example, special stickers are required for the collection of large household appliances, while the government of the Australian state of Victoria has taken an even stricter measure by banning the disposal of e-waste in landfills. The government combined this ban with educational programs for citizens and a financial package of AUD 16.5 million to support the safe disposal of e-waste [5]. In Hong Kong, a WEEE recycling facility called "WEEE Park" was developed by the government and started operations in 2017. The planned processing capacity is 30,000 tons per year [29]. Singapore has a well-established e-waste recycling program managed by the National Environment Agency (NEA). Electronic products can be dropped off at designated recycling bins or collection points located at various locations such as shopping malls, community centers, and electronics stores. These bins are usually marked with the "recycling" symbol or the NEA logo [30, 31]. **Figure 7** shows the legislation in other countries that are not located in Europe.

4.2 Pre-treatment

Pre-treatment is a necessary step in the recycling process, as the direct smelting of e-waste is not ideal due to its lower copper content compared to copper scrap. Dismantling, crushing, and pre-concentration are used to recover the materials and homogenize the feed for subsequent processes. Manual or semi-automatic dismantling and sorting processes are used to remove large, reusable, or hazardous components [33]. While dismantling is time-consuming, it offers a higher separation

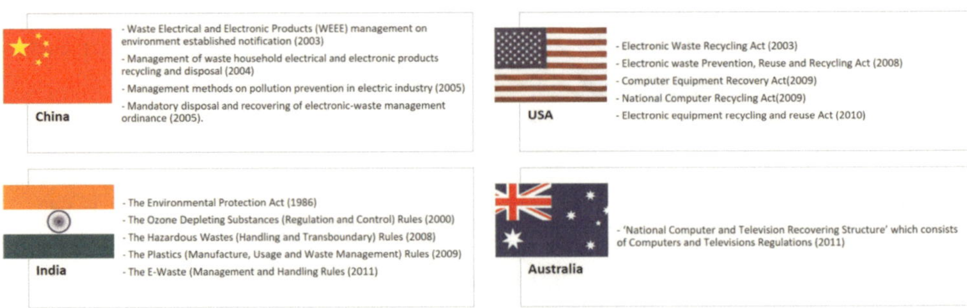

Figure 7.
E-waste management laws in other countries that are not located in Europe [32].

Crushing	Magnetic separation	Electrical separation	Gravity separation
Crushers (impact crusher, Hammer mill, etc.)	Cross belt magnetic separator	Corona-electrostatic separation	Sink-float separation
Shredding and cutting	Induced roll magnetic separator	Tribo-electrostatic separation	Jig separation
Grinding and pulverizing	High intensity magnetic separator	Eddy current separation	Shaking table

Table 1.
Common physical separation methods for pre-treatment of e-waste.

efficiency compared to other methods. The next step is crushing, which serves two important purposes: liberation and size reduction for easier handling, processing, and usability. Crushers and shredders are more efficient, as e-waste consists of both brittle materials (e.g., glass and ceramics) and tough materials (e.g., plastics and metals). The shredded materials then undergo a series of operations including screening/classification, magnetic/electrostatic separation, gravity separation, eddy current separation, and flotation to recover and separate ferrous alloys, aluminum scrap, copper-bearing scrap, and plastics [2]. The physical processes used in e-waste recycling often have low recovery efficiency and high energy consumption [13]. **Table 1** shows the common physical separation methods for pre-treatment of e-waste.

4.3 Pyrometallurgy

Pyrometallurgy is a process in which non-ferrous metals are refined or extracted from metallurgical materials at high temperatures. It includes processes such as incineration, melting, drossing, smelting, and roasting to extract valuable metals from e-waste [33]. The selection of specific pyrometallurgy processes depends on the type of e-waste and the requirements of the melting process [34]. Various metals, including Cu, Ag, Au, Pd, Ni, Se, Zn, and Pb, can be recovered from different sources of e-waste using this method. However, maintaining the operating parameters for successful recovery of precious metals can be challenging. One of the biggest challenges is balancing the heat and the materials to be treated to maintain the high temperatures [1]. In addition, predicting the mechanical aspects of the recovery process is complex as the metallurgical reactions quickly reach equilibrium, leading to changes in the chemical and phase composition of the e-waste. The pyrometallurgical process also produces by-products such as slag, soot, and flue gases, which consist of high-temperature dust, smoke, and toxic gases. These toxic gases, including dibenzo-p-dioxin, biphenyl, anthracene, polybrominated dibenzofurans (PBDFs), and polybrominated dibenzodioxins (PBDDs), can have adverse effects on human health and the atmosphere. The formation of slag can also affect the yield of the process by hindering metal recovery. There are a limited number of recycling plants using pyrometallurgical methods, such as the Ronnskar smelter in Sweden, Umicore in Belgium, Aurubis in Germany, and Noranda in Canada [35, 36]. Researchers are continuously working on improving the yield and reducing the emissions associated with this method.

After pre-concentration, the copper-bearing scrap is further processed in existing smelters to extract metal. Typically, copper smelters process enriched copper scrap as a secondary raw material in various plants, depending on the purity of the scrap.

The Swedish Boliden Ltd., for example, feeds scrap with a high copper content into the converting process, while scrap with a low copper content is fed into the Kaldo furnace [37]. In Japan, Dowa's Kosaka smelter has been using copper scrap from e-waste as a secondary feedstock in its Outokumpu flash smelter since 2002, and in 2006, around 10% of the plant's copper throughput came from e-waste [38]. Plastics in e-waste can be useful as an energy source and reducing agent for some smelters. For example, Umicore's IsaSmelt™ TSL (Top Submerged Lance) process has successfully used plastic-rich e-waste as a reducing agent instead of coke without any problems [37]. As the generation of e-waste continues to increase, specialized, stand-alone smelters may become necessary to process these materials. Flores et al. [39] proposed the construction of an IsaSmelt™ TSL plant that processes copper scrap from e-waste and copper wire waste, which could open up new opportunities in countries where this waste is currently exported for processing. Pyrometallurgical techniques such as incineration, pyrolysis, and combustion can also be used as thermal pre-treatment methods to produce the high copper scrap content required by smelters. Incineration, in which organic compounds are thermally destroyed to reduce volume, is considered an obsolete technique due to the release of unwanted CO_2, particulates/dust, and pollutants such as mercury and cadmium [40]. Pyrolysis, on the other hand, is a heat treatment process that is carried out without the presence of oxygen and produces organic end products such as oil and charcoal-metal-glass-ceramic mixtures that are easier to separate physically. Incineration, in which e-waste is heated to over 1000°C, is a cheaper alternative to incineration as it generates energy. Mitsubishi Materials' Naoshima smelter and refinery in Japan, for example, incinerates PCB-dominated e-waste to obtain slag-metal mixtures that are used as secondary feedstock in their patented Mitsubishi continuous smelting process [17]. Incineration is selected as a pre-treatment method to destroy brominated flame retardants (BFRs) in PCBs, which are bromine-containing hydrocarbons classified as persistent organic pollutants, and to avoid the release of problematic gaseous emissions such as dioxins during smelting. Pyrometallurgy is considered an ideal recycling strategy for e-waste in high-income countries with existing smelting facilities. However, it is not economically viable in low- and middle-income countries due to the significant capital expenditure, high operating costs, and advanced technical expertise required to build and maintain smelters [2].

4.4 Hydrometallurgy

Electronic waste (e-waste) contains a significant number of precious metals, making it a valuable resource. However, as electronic products become more environmentally friendly and contain fewer precious metals, research into hydrometallurgical methods has expanded beyond the recovery of precious metals to the recycling of precious and common metals [19].

This approach offers advantages such as low investment and high recovery rates, making it superior to pyrometallurgy [37]. Similar to the pyrometallurgical process, the hydrometallurgical process includes a pre-treatment step in which mechanical shredding is followed by a sorting process. This allows the metallic fractions to be exposed to chemical effects during the leaching process [35]. During leaching, solid materials react with chemical reagents called extractants or lixiviants, to extract the desired metals in a dispersion medium. Lixiviating agents such as halides, thiourea, thiosulfate, cyanide, HCl, H_2SO_4, and HNO3 are used to form metal complexes that facilitate the dissolution of the metallic components and their transfer to the leachate solution [1].

Hydrometallurgical technology for the recovery of metals from e-waste emerged in industrialized countries in the late 1960s. The basic principle is to expose the PCB scrap to acidic or alkaline leaching solutions, which separates the valuable metals from other materials. Subsequently, solvent extraction, electrowinning, precipitation, displacement, ion exchange, distillation, and filtration are used to recover precious and base metals from the liquid phase [19].

Hydrometallurgy is considered a promising alternative to pyrometallurgy for e-waste recycling, especially in low- and middle-income countries. It can be operated on a smaller scale and applied to low-grade waste [41]. In the past, the focus of e-waste hydrometallurgy has been on gold and silver, but there is growing interested in the recovery of other metals such as copper, nickel, and cobalt. For the recovery of gold and silver from e-waste, cyanide and aqua regia can be used, which are common lixiviants in the processing of gold ore. However, the abundance of copper and ferrous alloys in e-waste poses a problem related to reagent consumption and interference of Cu^{2+} during recovery [42]. One approach to address this problem is sulfuric acid leaching for the recovery of copper and iron alloys prior to gold extraction with cyanide or aqua regia [43]. However, the use of cyanide and aqua regia is becoming increasingly unpopular due to environmental concerns, so alternative leaching agents have been investigated since 2000. Copper, cobalt, and nickel can be extracted from e-waste using inorganic or organic solutions. Dilute nitric acid is effective for the oxidation of copper at room temperature, while dilute sulfuric acid is not suitable for copper dissolution without an oxidizing agent [44]. In the recovery phase, adsorption on activated carbon, cementation with aluminum, zero-valent iron or zinc, and electrowinning of metals are generally used.

In addition, technological advances, such as the use of ionic liquids as alternative solvents, have been investigated for the recycling of e-waste. Ionic liquids are salts that exist in a liquid state at low temperatures and have unique properties that make them attractive for metal extraction. They can selectively dissolve metals from e-waste and offer the potential for more efficient and environmentally friendly recycling processes. The Taiwanese company Uwin Nanotech has developed an automatic desoldering machine and a non-cyanide hydrometallurgical process for recovering gold, silver, palladium, copper, and tin from printed circuit boards. In recent developments, supercritical CO_2 has been identified as a promising lixiviant due to its ability to selectively break down the polymer matrix of PCBs, allowing easier separation [45]. On the other hand, in the field of e-waste pyrometallurgy, emphasis is placed on waste minimization through the reprocessing and reuse of slag/fly ash rather than making direct changes to processing methods [46]. The composition of e-waste is also expected to include a greater proportion of discarded renewable energy technologies such as photovoltaics (PV) [47] or electric vehicles (EV) [48], and research on these topics is expected to increase in the future.

In addition, research is being conducted into the development of integrated processes that combine different recycling methods to maximize metal recovery from e-waste. These integrated approaches aim to optimize the efficiency and sustainability of the recycling process by combining techniques such as mechanical processing, hydrometallurgy, and pyrometallurgy.

Leaching methods include mineral acids and ammonia-ammonium, where mineral acids include H_2SO_4, HNO_3, and HCl. Dilute acids are generally capable of dissolving base metals such as Zn, Sn, Fe, and Al, which are considered active metals. However, Cu must be leached with either an oxidizing acid or a non-oxidizing acid containing oxidizing agents. The concentration of acids is the most influential parameter during the leaching process [49].

4.4.1 Nitric acid

- Copper can be oxidized to copper nitrate with dilute nitric acid at normal temperature without an additional oxidizing agent.
- Leaching produces acid gas.
- Copper nitrate cannot be used directly for electroplating high-purity copper. It must be converted into a pure copper sulfate solution by extraction and back-extraction before electroplating.

$$3Cu + 8HNO_3\ (\text{dilute}) = 3Cu(NO_3)_2 + 2NO + 4H_2O \quad (1)$$

$$Cu + 4HNO_3\ (\text{concentrated}) = Cu(NO_3)_2 = 2NO_2 + 2H_2O \quad (2)$$

Zhang used two types of recycling leaching systems, H_2SO_4-HNO_3-H_2O-NaOH and H_2SO_4-HNO_3-H_2ONO_X, for the pre-treatment of electronic waste. Lix84I was used as the extracting agent, sulfuric acid was used as the extracting agent, and copper purity of more than 99.9% was obtained. The current efficiency is over 90% [19]. Hoang et al. [50] discovered in their study that a recycling efficiency of 99.7% for Cu could be achieved by using HNO3 at a concentration of 3.5 mol/L and LIX 984 N at a concentration of 50%, while maintaining an A/O phase ratio of 1:1.5.

4.4.2 Sulfuric acid

- Copper requires an additional oxidizing agent, and H_2SO_4 cannot dissolve it directly.
- Acid leaching with sulfuric acid has a short reaction time and high efficiency (99%).
- However, sulfuric acid has strong corrosive properties that require special reactor materials.
- Impurity ions present in the leaching process complicate the subsequent separation of the metal ions.

Chi et al. [51] used mechanical and physical methods to sort out the non-magnetic metals from the PCB waste. Then, they soaked the residue to leach copper, iron, zinc, nickel, and aluminum with sulfuric acid and hydrogen peroxide at 85°C. The leaching rate was higher than 95%, and the rest of the solid can be recovered by ammonium sulfate, copper sulfate, and ammonia. Zhang et al. [52] used sulfuric acid and hydrogen peroxide as reaction reagents for the leaching of copper and then used the electrolysis-electrodialysis method to recover copper from the leaching liquid; the copper recovery rate was 88.07%. The study conducted by Kumari et al. brought to light the correlation between increased pressure and improved metal dissolution. Under the conditions of 2.4 M H2SO4 at a temperature of 150°C for a duration of 90 minutes and an oxygen pressure of 20 bar, the recovery rate of Cu, Ni, and Fe was about 99%. To increase the leaching efficiency of Cu leaching, oxidizing agents such as $CuCl_2$ and H_2O_2 were commonly used. When using a 15% by-weight H_2SO_4 solution over a period of 3 hours, the addition of 10 ml of 30% H2O2 to the leaching agent resulted in the leaching of almost all of the Cu [53].

4.4.3 Hydrochloric acid

The recycling of copper by chloride hydrometallurgy offers numerous advantages over conventional processes for the regeneration of mixed copper. These advantages include shorter process time, simpler equipment requirements, lower capital investment, lower energy consumption, higher metal recovery rates, lower processing costs, and non-polluting environmental benefits. To facilitate the chloride leaching of copper in PCB waste, Yu [54] gave guidance on the leaching mechanism and technological process using copper chloride solutions (NaCl, HCl, $CuCl_2$) for recycling other copper. By using $CuCl_2$ as an oxidizing agent, the copper in the other copper was oxidized to Cu^+ ions, which then combined with Cl^- ions to form copper chloride complexes and finally enabled copper recovery by electrodeposition. The apparent activation energy of the leaching reaction indicated that the process of copper chloride salt leaching followed a typical diffusion-controlled mechanism. Masao [19] investigated the dissolution of copper in copper chloride hydrochloric acid solutions using dissolved oxygen molecules. The results showed that the concentration of the antioxidant $CuCl_2$ and the complexing agent Cl^- significantly influenced the dissolution rate of copper. Under the experimental conditions, the oxygen partial pressure and hydrogen ion concentration had no effect on the copper dissolution rate, and the reaction mechanism for copper dissolution was elucidated.

Fe^{3+} ions derived from $Fe_2(SO_4)_3$ and $FeCl_3$ exhibit strong oxidizing properties and are commonly used as oxidants in leaching processes. $Fe_2(SO_4)_3$ is not only easily accessible but can also provide Cl^- ions and form complexes with metals, accelerating the dissolution of metals. Acidulated solutions of $FeCl_3$ are often used for the treatment and recycling of metals. In a study conducted by Cakir [55], the effects of chemical etching of copper using $FeCl_3$ and $CuCl_2$ solutions were compared, and the results showed a higher etching rate when $FeCl_3$ was used as an etchant. This can be attributed to the use of ferric chloride as an oxidizing agent, which triggers the following chemical reactions:

$$FeCl_3 + Cu = FeCl_2 + CuCl \tag{3}$$

$$FeCl_3 + CuCl = FeCl_2 + CuCl_2 \tag{4}$$

$$CuCl_2 + Cu = 2CuCl \tag{5}$$

4.4.4 Ammonia-ammonium salt

- Ammonia or ammonium salt can react with copper in the presence of an oxidizing agent and form a copper-ammonia complex ion that dissolves the copper.

- The pH of the leaching system affects the reaction equilibrium and the leaching rate of copper.

- The copper-ammonia complex solution produced cannot be used directly for electroplating and must be converted into pure copper sulfate solution through extraction-reverse extraction technology:

$$Cu + H_2O_2 + 4NH_3 \cdot H_2O = Cu(NH_3)_4{}^{2+} + 2OH^- + 4H_2O \tag{6}$$

$$Cu + H_2O_2 + 4NH^{4+} = Cu(NH_3)_4{}^{2+} + 2H^+ + 2H_2O \tag{7}$$

$$Cu + Cu(NH_3)_4{}^{2+} = 2Cu(NH_3)^{2+} \tag{8}$$

$$2Cu(NH_3)^{2+} + H_2O_2 + 4NH_3 \cdot H_2O + 2H^+ = 2Cu(NH_3)_4{}^{2+} + 6H_2O \tag{9}$$

Since the middle of the twentieth century, numerous researchers have investigated the process of copper solubilization with liquid ammonia. In ammonia-ammonium salt systems, copper is present in the form of a copper-ammonia complex. The presence of the copper-ammonium complex and molecular oxygen in the solution serves as an oxidizing agent and facilitates the separation of copper from other metal components [56]. In their studies, Rania et al. used an 8% ammonia-0.5 M ammonium citrate solution with a liquid-to-solid ratio of 10:1, which resulted in a leaching of almost 98% of the copper. In addition, more than 96% Cu was leached with ammonia-ammonium sulfate, and a purity of 99.5% was achieved [53]. Furthermore, $(NH_4)_2CO_3$ was used to leach 70% of Zn from ICT waste and 100% of Cu from WPCBs, as reported in [57].

In addition to conventional methods, various innovative approaches to base metal leaching have been explored, such as supercritical liquids, ionic liquids, electrochemical leaching [58], and bioleaching. For example, Calgaro et al. [59] achieved a copper leaching efficiency of up to 88.79% by using a lixiviant consisting of supercritical CO_2 with 20% H_2O_2 and 2.5 M H_2SO_4 in a solid-liquid ratio of 1:2 within a 20-min time frame. In another study, it was reported that the recovery rate of copper could be increased from 57.64% to 90.94% by slurry electrolysis by replacing 10% H_2SO_4 with $[BSO_3HP_y]HSO_4$ [60]. Electrochemical methods have also been used for copper recovery, where an H_2SO_4-$CuSO_4$-NaCl system was successfully used to recover small-grained cathode copper, with the electrolyte being regenerable [61].

4.5 Biohydrometallurgy

Biohydrometallurgy is a process in which microorganisms are used to dissolve elements from solid materials so that they can be recovered by separation processes. The recovery of copper through biohydrometallurgy is successful and has attracted a great deal of attention. The technique has been used since pre-Roman times to extract copper, silver, and aluminum from mines in the Rio Tinto region of Spain. However, its commercial application was only observed about ten years ago in a mine in Thorsis.

Biohydrometallurgy is considered a promising alternative strategy for metal extraction due to its low energy consumption and minimal use of chemicals. It can be operated in two different ways: single-stage and two-stage. In single-stage biohydrometallurgy, microorganisms and metals are combined with a nutrient medium, which enables microbial growth and subsequent dissolution of the metals in the solution. However, high concentrations of metal ions can be toxic to the microbes, restricting their growth and reducing the efficiency of recovery.

In two-stage biohydrometallurgy, the microbes are cultivated in a nutrient medium until they reach the stationary phase of their growth period. The waste material to be recycled is then added aseptically to initiate the recovery process.

Various organisms are used in the leaching of biohydrometallurgy, including *Acidobacillus thiooxidans*, *Acidobacillus ferrooxidans*, *Sulfolobus* sp., *Pseudomonas* sp., *Bacillus* sp., *Aspergillus* sp., and *Penicillium* sp. These organisms obtain energy from oxidation reactions or the reduction of sulfur-containing compounds. The mechanisms of metal transport include acidolysis, metal complexation, redox reactions, or bioaccumulation [1]. Bioleaching bacteria commonly used in the recycling process include acidophilic bacteria isolated from nature [62]. In the bioleaching process, these acidophilic bacteria use ions as electron donors. Thus, acidophilic bacteria such as Thiobacillus can efficiently leach over 90% of copper from PCBs. A similar phenomenon has been observed in acidophilic bacteria such as *Thiobacillus ferrooxidans*. A low concentration of PCBs is required for the recycling of PCBs on an industrial scale [63]. Fungi, such as *Aspergillus* and *Penicillium*, can also be used as leaching agents in the bioleaching process. One example is the fungal strain Y5 isolated from an industrial environment contaminated with heavy metals, in particular *Penicillium chrysogenum*. *Penicillium chrysogenum* was used in the recycling of copper-containing PCB waste, and its bioleaching efficiency was tested with different pulp densities [64].

Several studies (**Table 2**) have demonstrated the effectiveness of biohydrometallurgy in recovering copper from electronic waste (e-waste). The recovery efficiency can be over 80% or even 90%, depending on the organisms and conditions used. Factors such as the choice of microorganisms, the surface area, the pulp density, the choice of starting materials, the temperature, and the pH value all influence the efficiency of the biohydrometallurgical process (**Figure 8**).

Challenges with biohydrometallurgy include the long operating time and potential toxic effects on the microbes that need to be addressed to scale up the process to commercialization. The use of adaptable microorganisms that can tolerate harsh environmental conditions and the elucidation of the inner mechanisms of

Microorganism	Copper recovery (%)	Operational condition	Reference
Thiobacillus thiooxidans, Thiobacillus ferrooxidans	>90	Pulp density 5–15 g/l	[65]
Aspergillus niger	89	Pulp density of 10 g/l	[66]
Acidithiobacillus ferruoxidans	99	Concentration of Iron(III) ions 6.66 g/l	[67]
Aspergillus niger, Acidithiobacillus thiooxidans	83	Incubation 6 h, Tamperature 70–90 C	[68]
Pseudomonas chlororaphis	52.3	pH, temperature, cyanide concentration	[69]
Leptospirillum ferriphilm	100	Pulp density > 5%	[70]
Acidithiobacillus ferrooxidans	10	pH, oxidation reduction potential (ORP), leaching rate	[71]
Alicyclobacillus sp. and *Sulfobacillus* sp.	80	Biochar dose	[72]
Acinetobacter sp. CrB2	23	—	[73]
Acidithiobacillus ferrooxidans	74	Pulp density 10 g/l, pH 2.5, ferrous sulfate 40 g/l	[74]
Acidithiobacillus ferrooxidans	99	Pulp density 10–30 g, pH 2.0, temperature 303 K	[75]
Mixed microbial consortia iron and sulfur-oxidizing microorganisms	98	Pulp density 7–15%	[76]
Frankia casuarinae Frankia sp	94	Low pulp densities of 0.2%	[77]

Table 2.
Copper recovery from e-waste by biohydrometallurgical processes.

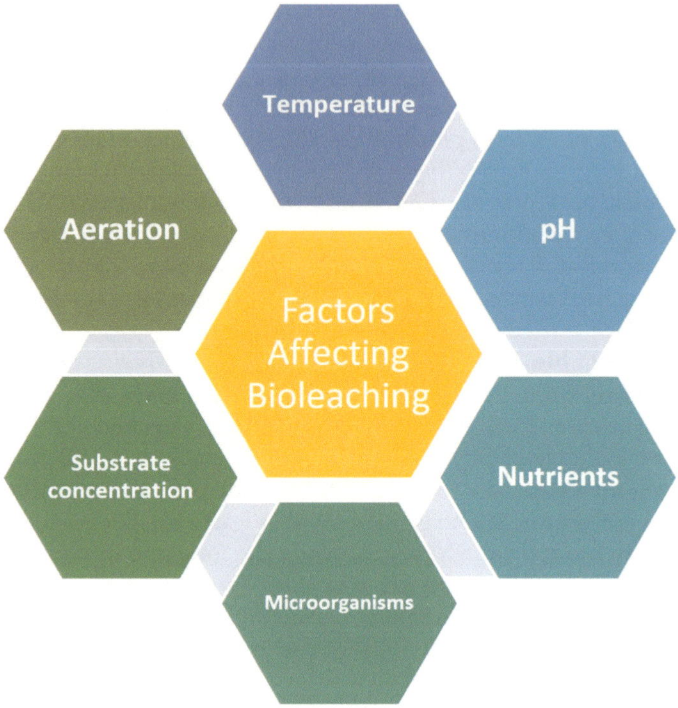

Figure 8.
Parameters that affect bioleaching [17].

biohydrometallurgy are areas that require further research. In addition, the use of continuous or modified reactors for the recovery of e-waste should be systematically investigated in the future [1]. The bioleaching process of copper (Cu) extracted from PCBs by *A. ferrooxidans* follows an indirect leaching mechanism facilitated by biogenic sulfuric acid, similar to metal sulfides. The microorganism oxidizes sulfur, which leads to the production of sulfuric acid. Iron(II) ions and sulfur are introduced into the leaching medium as electron donors. The presence of iron(II) ions acts as an oxidizing agent and thus accelerates the leaching process. Copper is mobilized by the interaction of iron(III) ions and sulfuric acid, which leads to the acidolysis-redox-lysis-bioleaching mechanism for metal dissolution. The concentration of ferric ions is critical for the leaching rate and extraction efficiency. A favorable association between PCB particles and *A. ferrooxidans* occurs when the attractive van der Waals force outweighs the repulsive force [78].

5. Conclusions

The challenges associated with the exploitation of resources go beyond the depletion of reserves. Difficult access to ores in low- and middle-income countries and the negative social perception of mining and metal processing are significant obstacles. Overcoming these challenges requires a strategic approach that also takes socio-cultural aspects into account. This includes engaging communities and raising awareness of the benefits of mining as well as combating negative perceptions through education and effective communication. By incorporating socio-cultural considerations, it becomes possible to promote sustainable resource management and ensure responsible extraction of copper and critical metals.

The increasing demand for copper and critical metals, driven by the need for renewable energy and clean technologies, poses a challenge as natural deposits of these metals are at risk of rapid depletion without efficient recycling strategies. To extend the lifespan of reserves, it is crucial to improve recycling efficiency, particularly through greater household participation in waste separation and the implementation of supportive government legislation. In this way, the depletion of copper reserves can be significantly delayed, and a sustainable supply of these important metals can be ensured for the next 30 years and beyond.

To improve the recovery and recycling rates of copper and critical metals, the development of better processing technologies and environmentally friendly techniques is essential. This includes specialized facilities for e-waste and the introduction of sustainable mine waste management systems. In addition, new issues such as the removal of microplastics in recycled wastewater need to be addressed. By continuously developing processing technologies and introducing sustainable practices, it is possible to maximize the benefits from resource use while minimizing the environmental impact.

Pyrometallurgy and hydrometallurgy are two well-known methods used in the recycling of electronic waste (e-waste) to recover valuable metals. Pyrometallurgy involves high-temperature processes such as incineration, melting, smelting, and roasting to refine or extract non-ferrous metals such as copper, silver, gold, palladium, nickel, selenium, zinc, and lead from e-waste. Pyrometallurgy is used in recycling plants such as the Ronnskar smelter in Sweden and Umicore in Belgium, but the challenges lie in balancing heat and material, maintaining high temperatures, and dealing with the environmental impact of the resulting by-products such as slag and toxic gases. Efforts are being made to improve yields and reduce emissions in pyrometallurgical processes. Hydrometallurgy, on the other hand, offers a promising alternative to pyrometallurgy, especially in low- and middle-income countries where smaller operations and the treatment of low-grade waste are more feasible. This method involves a pre-treatment step followed by leaching with chemical reagents to extract metals from the e-waste. Subsequent recovery processes such as solvent extraction and electrowinning are used to separate and recover the extracted metals. Cyanide and aqua regia are traditionally used for the recovery of gold and silver in hydrometallurgy, but alternative leaching agents are being explored due to environmental concerns. In addition, technological advances such as the use of ionic liquids and supercritical CO_2 are being investigated to develop more efficient and environmentally friendly recycling processes for e-waste. Integrated approaches are also being developed that combine different recycling methods to maximize metal recovery and minimize waste generation. Both pyrometallurgy and hydrometallurgy play a crucial role in e-waste recycling. Ongoing research and development work is focused on improving their efficiency, sustainability, and environmental impact.

Biohydrometallurgy offers a promising approach to metal recovery, especially for copper. This process, whose history dates back to pre-Roman times in the Rio Tinto region of Spain, uses microorganisms to dissolve elements from solid materials. Biohydrometallurgy is characterized by its low energy consumption and minimal use of chemicals. It can be operated in either a one-stage or two-stage process, in which certain microbes are cultivated in a nutrient medium before the waste material is fed for recovery. Various organisms, including bacteria such as Acidobacillus and Thiobacillus and fungi such as Aspergillus and Penicillium, play an essential role in the biohydrometallurgical process, utilizing mechanisms such as acidolysis, metal complexation, redox reactions or bioaccumulation for metal transport.

6. Future prospectives

The future of e-waste recycling has significant implications for environmental sustainability and resource conservation. A combination of legislative measures, technological advances, circular economy principles, international cooperation, and public engagement is expected to shape the landscape of e-waste management in the coming years. Governments worldwide are recognizing the urgent need to address the growing e-waste crisis and are expected to enact stricter laws and regulations to ensure proper disposal and recycling practices. For example, extended producer responsibility (EPR) regulations, which make manufacturers responsible for the disposal of their products at the end of their life, are set to take hold.

In parallel, ongoing research and development efforts in the field of e-waste recycling will focus on the further development of innovative technologies. These technologies aim to improve the efficiency and sustainability of e-waste recycling processes. Advanced separation techniques such as hydrometallurgical and biotechnological processes promise more effective extraction of valuable metals from e-waste, reducing reliance on traditional smelting processes.

The future of e-waste recycling is also expected to see a major shift towards a circular economy approach. This approach emphasizes the recyclability of electronic devices by promoting modular designs and environmentally friendly materials. By recovering valuable resources from e-waste, the circular economy aims to minimize the amount of waste and reduce the need to extract raw materials.

However, effective recycling of e-waste requires more than just legislation and technological advances. Raising public awareness and participation are essential components of a sustainable e-waste management strategy. Education campaigns targeting consumers should raise awareness about the environmental and health impacts of e-waste and encourage responsible disposal practices. Additionally, the establishment of convenient collection points and incentivized return programs can encourage citizens to actively participate in e-waste recycling initiatives.

Author details

Hadi Sharifidarabad
School of Metallurgy and Materials Engineering, Iran University of Science and Technology, Tehran, Iran

*Address all correspondence to: hadisharifi@metaleng.iust.ac.ir

References

[1] Nithya R, Sivasankari C, Thirunavukkarasu A. Electronic waste generation, regulation and metal recovery: A review. Environmental Chemistry Letters. 2021;**19**: 1347-1368

[2] Tabelin CB, Park I, Phengsaart T, Jeon S, Villacorte-Tabelin M, Alonzo D, et al. Copper and critical metals production from porphyry ores and E-wastes: A review of resource availability, processing/recycling challenges, socio-environmental aspects, and sustainability issues. Resources, Conservation and Recycling. 2021;**170**: 105610

[3] Baldé CP, Forti V, Gray V, Kuehr R, Stegmann P. The Global E-waste Monitor 2017: Quantities, Flows and Resources. Bonn, Geneva and Vienna: United Nations University (UNU), International Telecommunication Union (ITU), and International Solid Waste Association (ISA); 2017

[4] Srivastava RR, Pathak P. Policy issues for efficient management of E-waste in developing countries. In: Handbook of Electronic Waste Management. Butterworth-Heinemann: Elsevier; 2020. pp. 81-99

[5] Forti V, Balde CP, Kuehr R, Bel G. The Global E-waste Monitor 2020: Quantities, Flows and the Circular Economy Potential. Bonn, Geneva and Vienna: United Nations University/United Nations Institute for Training and Research, International Telecommunication Union, and International Solid Waste Association; 2020

[6] Alam T, Golmohammadzadeh R, Faraji F, Shahabuddin M. E-waste recycling technologies: An overview, challenges and future perspectives. In: Paradigm Shift in E-Waste Management. 1st Edition. CRC Press; 2022. pp. 143-176

[7] Masud MH, Mourshed M, Mahjabeen M, Ananno AA, Dabnichki P. Global electronic waste management: Current status and way forward. In: Paradigm Shift in E-Waste Management. 1st Edition. CRC Press; 2022. pp. 9-47

[8] Borthakur A, Govind M. Public understandings of E-waste and its disposal in urban India: From a review towards a conceptual framework. Journal of Cleaner Production. 2018;**172**: 1053-1066

[9] Borthakur A, Govind M, Singh P. Inventorization of E-Waste and its Disposal Practices with Benchmarks for Depollution: The Global Scenario. In: Electronic Waste Management and Treatment Technology. Butterworth-Heinemann: Elsevier; 2019. pp. 35-52

[10] Yaashikaa P, Priyanka B, Kumar PS, Karishma S, Jeevanantham S, Indraganti S. A review on recent advancements in recovery of valuable and toxic metals from e-waste using bioleaching approach. Chemosphere. 2022;**287**:132230

[11] Cucchiella F, D'Adamo I, Koh SL, Rosa P. Recycling of WEEEs: An economic assessment of present and future e-waste streams. Renewable and Sustainable Energy Reviews. 2015;**51**: 263-272

[12] Yang J, Wang H, Zhang G, Bai X, Zhao X, He Y. Recycling organics from non-metallic fraction of waste printed circuit boards by a novel conical surface triboelectric separator. Resources, Conservation and Recycling. 2019;**146**: 264-269

[13] Hao J, Wang Y, Wu Y, Guo F. Metal recovery from waste printed circuit boards: A review for current status and perspectives. Resources, Conservation and Recycling. 2020;**157**: 104787

[14] Kasper AC, Bernardes AM, Veit HM. Characterization and recovery of polymers from mobile phone scrap. Waste Management & Research. 2011;**29**(7):714-726

[15] Chen Y, Chen M, Li Y, Wang B, Chen S, Xu Z. Impact of technological innovation and regulation development on e-waste toxicity: A case study of waste mobile phones. Scientific Reports. 2018;**8**(1):7100

[16] Needhidasan S, Samuel M, Chidambaram R. Electronic waste–an emerging threat to the environment of urban India. Journal of Environmental Health Science and Engineering. 2014; **12**:1-9

[17] Tokyo 2020 Medal Project. Towards an Innovative Future for All [Internet]. Tokyo. 2020. Available from: https://tokyo2020.org/en/games/medals-project/

[18] Wang J, Xu Z. Disposing and recycling waste printed circuit boards: Disconnecting, resource recovery, and pollution control. Environmental Science & Technology. 2015;**49**(2):721-733

[19] Xu Y, Li J, Liu L. Current status and future perspective of recycling copper by hydrometallurgy from waste printed circuit boards. Procedia Environmental Sciences. 2016;**31**:162-170

[20] Khaliq A, Rhamdhani MA, Brooks G, Masood S. Metal extraction processes for electronic waste and existing industrial routes: A review and Australian perspective. Resources. 2014;**3**(1):152-179

[21] Fu J, Zhang A, Wang T, Qu G, Shao J, Yuan B, et al. Influence of e-waste dismantling and its regulations: Temporal trend, spatial distribution of heavy metals in rice grains, and its potential health risk. Environmental Science & Technology. 2013;**47**(13):7437-7445

[22] Song M, Luo C, Li F, Jiang L, Wang Y, Zhang D, et al. Anaerobic degradation of polychlorinated biphenyls (PCBs) and polychlorinated biphenyls ethers (PBDEs), and microbial community dynamics of electronic waste-contaminated soil. Science of the Total Environment. 2015;**502**:426-433

[23] Matsukami H, Tue NM, Suzuki G, Someya M, Viet PH, Takahashi S, et al. Flame retardant emission from e-waste recycling operation in northern Vietnam: Environmental occurrence of emerging organophosphorus esters used as alternatives for PBDEs. Science of the Total Environment. 2015;**514**:492-499

[24] Cayumil R, Khanna R, Ikram-Ul-Haq M, Rajarao R, Hill A, Sahajwalla V. Generation of copper rich metallic phases from waste printed circuit boards. Waste Management. 2014; **34**(10):1783-1792

[25] Ding Y, Zhang S, Liu B, Zheng H, Chang C-c, Ekberg C. Recovery of precious metals from electronic waste and spent catalysts: A review. Resources, Conservation and Recycling. 2019;**141**: 284-298

[26] Vakilchap F, Mousavi S, Shojaosadati S. Role of *Aspergillus niger* in recovery enhancement of valuable metals from produced red mud in Bayer process. Bioresource Technology. 2016; **218**:991-998

[27] Sahajwalla V, Hossain R. The science of microrecycling: A review of selective

synthesis of materials from electronic waste. Materials Today Sustainability. 2020;**9**:100040

[28] Forti V, Baldé K, Kuehr R. E-Waste Statistics: Guidelines on Classifications, Reporting and Indicators. Bonn: United Nations University; 2018

[29] E-waste Collection. Waste Reduction [Internet]. Available from: www.wastereduction.gov.hk, https://www.wastereduction.gov.hk/en-hk/waste-reduction-programme/e-waste-collection

[30] Where to Recycle E-Waste [Internet]. Available from: www.nea.gov.sg, https://www.nea.gov.sg/our-services/waste-management/3r-programmes-and-resources/e-waste-management/where-to-recycle-e-waste

[31] Where to Safely Recycle your Electronic Waste [Internet] [cited 2024 Mar 18]. Available from: www.nimbushomes.com, https://www.nimbushomes.com/blog/where-to-recycle-electronic-waste

[32] Islam A, Ahmed T, Awual R, Rahman A, Sultana M, Aziz A, et al. Advances in sustainable approaches to recover metals from e-waste: A review. Journal of Cleaner Production. 2020;**244**:118815

[33] Kaya M. Recovery of metals and nonmetals from electronic waste by physical and chemical recycling processes. Waste Management. 2016;**57**:64-90

[34] Ma E. Recovery of waste printed circuit boards through pyrometallurgy. In: Electronic Waste Management and Treatment Technology, . Butterworth-Heinemann, Elsevier; 2019. pp. 247-267

[35] Priya A, Hait S. Comparative assessment of metallurgical recovery of metals from electronic waste with special emphasis on bioleaching. Environmental Science and Pollution Research. 2017;**24**:6989-7008

[36] Priya J, Randhawa N, Hait J, Bordoloi N, Patel J. High-purity copper recycled from smelter dust by sulfation roasting, water leaching and electrorefining. Environmental Chemistry Letters. 2020;**18**:2133-2139

[37] Cui J, Zhang L. Metallurgical recovery of metals from electronic waste: A review. Journal of Hazardous Materials. 2008;**158**(2–3):228-256

[38] Hoshikawa Y. Copper smelting and refining at Kosaka smelter. Journal of MMIJ. 2007;**123**(12):597-601

[39] Alvear Flores GR, Nikolic S, Mackey PJ. ISASMELT™ for the recycling of E-scrap and copper in the US case study example of a new compact recycling plant. Journal of Metals. 2014;**66**(5):823-832

[40] Wang H, Zhang S, Li B, Pan Da WY, Zuo T. Recovery of waste printed circuit boards through pyrometallurgical processing: A review. Resources, Conservation and Recycling. 2017;**126**:209-218

[41] Akcil A, Erust C, Gahan CS, Ozgun M, Sahin M, Tuncuk A. Precious metal recovery from waste printed circuit boards using cyanide and non-cyanide lixiviants: A review. Waste Management. 2015;**45**:258-271

[42] Jeffrey MI, Ritchie IM. The leaching of gold in cyanide solutions in the presence of impurities II. The effect of silver. Journal of the Electrochemical Society. 2000;**147**(9):3272

[43] Kamberović Ž, Ranitović M, Korać M, Jovanović N, Tomović B, Gajić N. Pyro-refining of mechanically treated waste printed circuit boards in a DC arc-furnace. Journal of Sustainable Metallurgy. 2018;**4**:251-259

[44] Park I, Yoo K, Alorro RD, Kim M-s, Kim S-k. Leaching of copper from cuprous oxide in aerated sulfuric acid. Materials Transactions. 2017;**58**(10): 1500-1504

[45] Hsu E, Durning CJ, West AC, Park A-HA. Enhanced extraction of copper from electronic waste via induced morphological changes using supercritical CO2. Resources, Conservation and Recycling. 2021;**168**: 105296

[46] Panda R, Jadhao PR, Pant KK, Naik SN, Bhaskar T. Eco-friendly recovery of metals from waste mobile printed circuit boards using low temperature roasting. Journal of Hazardous Materials. 2020;**395**:122642

[47] Nain P, Kumar A. Initial metal contents and leaching rate constants of metals leached from end-of-life solar photovoltaic waste: An integrative literature review and analysis. Renewable and Sustainable Energy Reviews. 2020;**119**:109592

[48] Wang L, Wang X, Yang W. Optimal design of electric vehicle battery recycling network–from the perspective of electric vehicle manufacturers. Applied Energy. 2020;**275**:115328

[49] Bas AD, Deveci H, Yazici EY. Treatment of manufacturing scrap TV boards by nitric acid leaching. Separation and Purification Technology. 2014;**130**: 151-159

[50] Long Le H, Jeong J, Lee J-C, Pandey BD, Yoo J-M, Huyunh TH. Hydrometallurgical process for copper recovery from waste printed circuit boards (PCBs). Mineral Processing & Extractive Metallurgy Review. 2011; **32**(2):90-104

[51] Oh CJ, Lee SO, Yang HS, Ha TJ, Kim MJ. Selective leaching of valuable metals from waste printed circuit boards. Journal of the Air & Waste Management Association. 2003;**53**(7):897-902

[52] Zhang Z, Zhou L. Study on recovering copper from waste printed circuits plate. Liaoning Chemical Industry. 2005;**34**(3):93-95

[53] Yang H, Liu J, Yang J. Leaching copper from shredded particles of waste printed circuit boards. Journal of Hazardous Materials. 2011;**187**(1–3): 393-400

[54] Yu Z, Zhou B, Lu Z. Hydro-electro metallurgical process for recovering scrap from cuprous chloride solution. Shanghai Nonferrous Metals. 1999;**2**: 24-28

[55] Cakir O. Copper etching with cupric chloride and regeneration of waste etchant. Journal of Materials Processing Technology. 2006;**175**(1–3):63-68

[56] Rudnik E, Pierzynka M, Handzlik P. Ammoniacal leaching and recovery of copper from alloyed low-grade e-waste. Journal of Material Cycles and Waste Management. 2016;**18**: 318-328

[57] Wang M, Cao H-B, Zhang Y. Selective recovery of copper, zinc and nickel from printed circuit boards by ammonia leaching under pressure. Huan jing ke xue= Huanjing kexue. 2011;**32**(2): 596-602

[58] Ping Z, ZeYun F, Jie L, Qiang L, GuangRen Q, Ming Z.

Enhancement of leaching copper by electro-oxidation from metal powders of waste printed circuit board. Journal of Hazardous Materials. 2009;**166**(2–3): 746-750

[59] Calgaro CO, Schlemmer DF, Da Silva M, Maziero EV, Tanabe EH, Bertuol DA. Fast copper extraction from printed circuit boards using supercritical carbon dioxide. Waste Management. 2015;**45**:289-297

[60] Zhang G, He Y, Feng Y, Zhang T, Wang H, Zhu X. Recovery of residual metals from fine nonmetallic fractions of waste printed circuit boards using a vibrated gas-solid fluidized bed. Separation and Purification Technology. 2018;**207**:321-328

[61] Cocchiara C, Dorneanu S-A, Inguanta R, Sunseri C, Ilea P. Dismantling and electrochemical copper recovery from waste printed circuit boards in H_2SO_4–$CuSO_4$–NaCl solutions. Journal of Cleaner Production. 2019; **230**:170-179

[62] Fathollahzadeh H, Khaleque HN, Eksteen J, Kaksonen AH, Watkin EL. Effect of glycine on bioleaching of rare earth elements from Western Australian monazite by heterotrophic and autotrophic microorganisms. Hydrometallurgy. 2019;**189**:105137

[63] Xia M-C, Wang Y-P, Peng T-J, Shen L, Yu R-L, Liu Y-D, et al. Recycling of metals from pretreated waste printed circuit boards effectively in stirred tank reactor by a moderately thermophilic culture. Journal of Bioscience and Bioengineering. 2017; **123**(6):714-721

[64] Xia M-C, Bao P, Liu A-J, Zhang S-S, Peng T-J, Shen L, et al. Isolation and identification of *Penicillium chrysogenum* strain Y5 and its copper extraction characterization from waste printed circuit boards. Journal of Bioscience and Bioengineering. 2018;**126**(1):78-87

[65] Brandl H, Bosshard R, Wegmann M. Computer-munching microbes: Metal leaching from electronic scrap by bacteria and fungi. Hydrometallurgy. 2001;**59**(2–3):319-326

[66] Ilyas S, Anwar MA, Niazi SB, Ghauri MA. Bioleaching of metals from electronic scrap by moderately thermophilic acidophilic bacteria. Hydrometallurgy. 2007;**88**(1–4):180-188

[67] Yang T, Xu Z, Wen J, Yang L. Factors influencing bioleaching copper from waste printed circuit boards by *Acidithiobacillus* ferrooxidans. Hydrometallurgy. 2009;**97**(1–2):29-32

[68] Saidan M, Brown B, Valix M. Leaching of electronic waste using biometabolised acids. Chinese Journal of Chemical Engineering. 2012;**20**(3): 530-534

[69] Ruan J, Zhu X, Qian Y, Hu J. A new strain for recovering precious metals from waste printed circuit boards. Waste Management. 2014;**34**(5):901-907

[70] Bryan C, Watkin E, McCredden T, Wong Z, Harrison S, Kaksonen A. The use of pyrite as a source of lixiviant in the bioleaching of electronic waste. Hydrometallurgy. 2015;**152**:33-43

[71] Bai J, Gu W, Dai J, Zhang C, Yuan W, Deng M, et al. The catalytic role of nitrogen-doped carbon nanotubes in bioleaching copper from waste printed circuit boards. Polish Journal of Environmental Studies. 2016;**25**(3): 951-957

[72] Wang S, Zheng Y, Yan W, Chen L, Mahadevan GD, Zhao F. Enhanced

bioleaching efficiency of metals from E-wastes driven by biochar. Journal of Hazardous Materials. 2016;**320**:393-400

[73] Jagannath A, Shetty V, Saidutta M. Bioleaching of copper from electronic waste using Acinetobacter sp. Cr B2 in a pulsed plate column operated in batch and sequential batch mode. Journal of Environmental Chemical Engineering. 2017;**5**(2):1599-1607

[74] Gu W, Bai J, Dong B, Zhuang X, Zhao J, Zhang C, et al. Enhanced bioleaching efficiency of copper from waste printed circuit board driven by nitrogen-doped carbon nanotubes modified electrode. Chemical Engineering Journal. 2017;**324**:122-129

[75] Nithya R, Sivasankari C, Thirunavukkarasu A, Selvasembian R. Novel adsorbent prepared from bio-hydrometallurgical leachate from waste printed circuit board used for the removal of methylene blue from aqueous solution. Microchemical Journal. 2018; **142**:321-328

[76] Garg H, Nagar N, Ellamparuthy G, Angadi SI, Gahan CS. Bench scale microbial catalysed leaching of mobile phone PCBs with an increasing pulp density. Heliyon. 2019;**5**(12)e02883:1-11

[77] Marappa N, Ramachandran L, Dharumadurai D, Nooruddin T. Recovery of gold and other precious metal resources from environmental polluted E-waste printed circuit board by bioleaching Frankia. International Journal of Environmental Research. 2020;**14**:165-176

[78] Kasper AC, Veit HM. Gold recovery from printed circuit boards of mobile phones scraps using a leaching solution alternative to cyanide. Brazilian Journal of Chemical Engineering. 2018;**35**: 931-942